P9-DDG-526

His thoughts go this way and that without order, reality and dreams, wishes and events that actually happened, place and time pass his soul in foglike rolling motion; most of the time he does not know where he is, believes himself to be traveling, in America, on the boat . . . It is always as if his soul were far, far away, in a beautiful, noble sphere where only science and eternal laws rule, and then that does not correspond to anything which surrounds him and he becomes unclear and confused.

—Anna von Helmholtz to her sister, 18 July 1894

NIGHT THOUGHTS *of a* CLASSICAL PHYSICIST

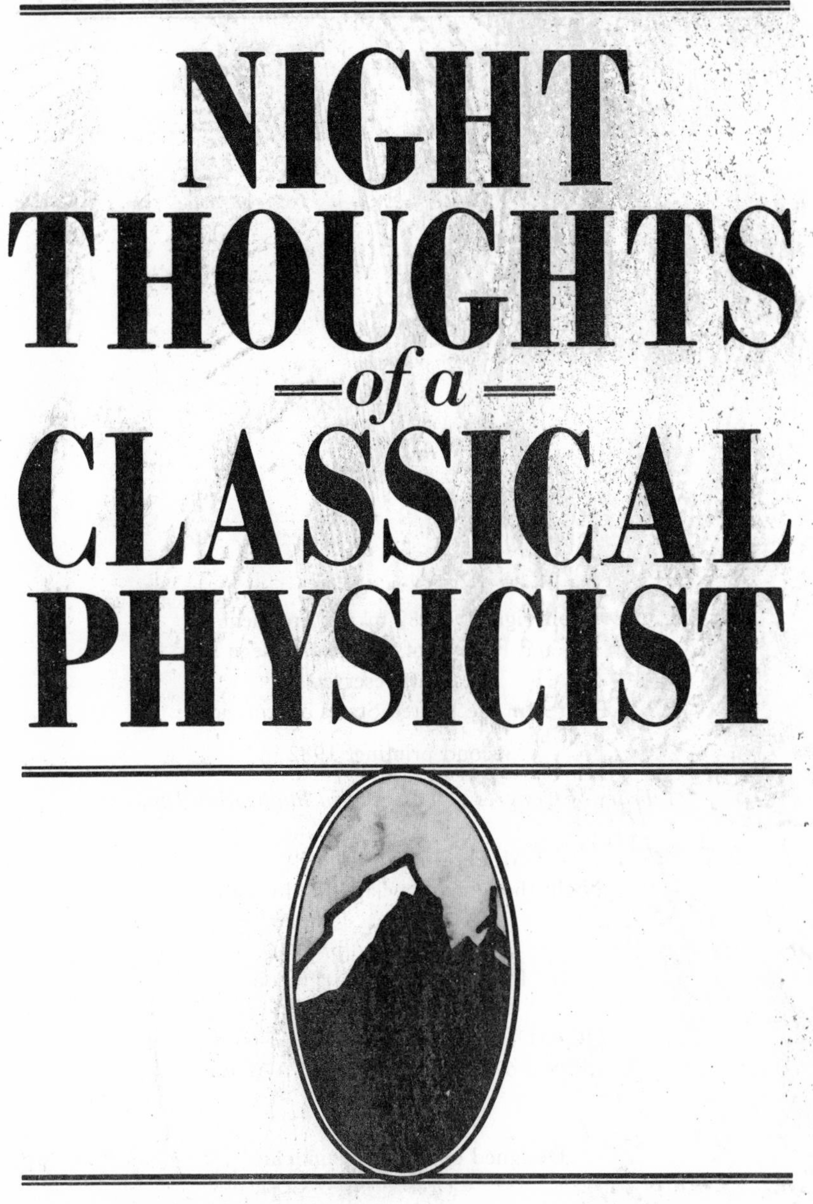

Russell McCormmach

Harvard University Press · Cambridge, Massachusetts · London, England · 1982

Printed in the United States of America

Second printing, 1982

Library of Congress Cataloging in Publication Data

McCormmach, Russell.
Night thoughts of a classical physicist.

Bibliography: p.
1. Physics—History—Fiction.
I. Title.
QC7.M35 813'.54 81-6674
ISBN 0-674-62460-2 AACR2

Designed by Gwen Frankfeldt

For Christa Jungnickel

CONTENTS

ILLUSTRATIONS

THE TOWN

WOULD IT EVER GET LIGHT, THE OLD PROFESSOR WONdered. He felt the cold rise up his arms.

Through the chorale the old professor had sat unmoving, and resisting the warm feeling musically uniting the audience. He couldn't rise to the passion of the music. He resented this blend of German patriotism with the current cult of Bach, and anyway he had heard enough of the emotional work of the young Bach. He liked the more lawful forms of the mature Bach, the classical structures built upon fundamental principles. They reminded him of his own physics . . . But not this. Over the monotonously beating organ, voices were laying down a restless, nibbling complaint of sorrow. Then came the final cantata, a timely celebration of victory through heavenly power. The fall of the enemy was prefigured in the song of joyous battle.

A severe headache was bringing the old professor closer to tears than any music would. Fortunately, the planners of tonight's program had recognized that the hearers were the nervous people of the twentieth century and accordingly shortened the musical prelude. He applauded vigorously for a few moments. He didn't look at the audience, but busied himself with the papers in his lap. Someone seemed to have left open a stagedoor to the night air. In his evening dress he was shivering.

He clenched the invitation, which read stiffly: You have the opportunity to be active in a patriotic manner. He knew that the university town was just small enough to make this event an obligation as well as an opportunity. And it was just large enough that he didn't recognize many in the audience apart from members of his circle, the prorector and senate from the university and the Goethe Society.

PATRIOTIC EVENINGS. HERR PROFESSOR DR. VICTOR JAKOB WILL SPEAK ON HIS EXPERIENCES OF THE YEAR 1870.

But it wasn't his turn yet, as imperial flags with golden fringes were moving toward him down the center aisle. The prorector bent toward the professor and pointed to an insertion in his invitation. ARTISTIC RENDERINGS OF PATRIOTIC IDEAS. Would the evening never end!

From previous evenings like this one, he remembered a talk on culture and war and another on the blessings of war. He had missed the talk on the moral and physical characteristics of soldiers, but not the one on the significance of the Prussians in this war, which proved most popular. His own talk was different in that it was a philosophical reflection. It was hard for him, since public reticence on personal matters was the sine qua non of his science. It was hard, too, since he had to keep the German destiny before his audience's eyes, which meant in a certain sense identifying himself with it.

A boy's soprano rose: I had a comrade at arms, you will not find a better . . . Belts held in place uniforms that were too large for these children of, what, thirteen or fourteen? They would soon grow into them and wear them in earnest. Standing in a semicircle, they lowered their flags over a boy on the floor. Another boy knelt beside the fallen and held up his rifle in a gesture of victory. The young voice went on solemnly: The drum called us to battle.

Now Professor Jakob read aloud from his carefully composed recollections: I was a junior officer when Germany fought France alone. From that war between two peoples, a united Germany arose. From today's world war, a Germany will arise that we cannot foresee. Still, with the worst possible outcome—loss of wealth, colonies, and provinces—we cannot lose the national feeling that we won on the battlefields of France in 1870 and 1871. Even if the enemies of Germany try to divide what Bismarck joined, they will not succeed in the long run. With our national feeling, Germany will rise to be a world power again.

Those thoughts were well received. Yet he couldn't go on, unable to focus on the words in front of his eyes. So he looked up from the lecture and began to tell a story. He and his platoon had been crawling up a wooded hill looking for a place to shoot from. Shells and grenades were falling all around them, and shrapnel and flint balls were flying just over their heads. Yet he (stupidly!) stood up and walked out of the woods to see the enemy better. It had been then that . . . A sharp pain in the side of his head drove the image out of his mind. Professor Jakob grasped the lectern and murmured an apology. Perhaps we should now turn to the young for tales of heroism, not to the old.

There was applause, though he scarcely heard it. He sat down with the prorector's discreet support. On that field of slaughter fifty years before, his head had been grazed by a bullet, but that now seemed nothing compared to the explosion of pain in his head a moment ago. One more musical act of communal uplift to endure, and he could go home. The audience sang: Dear Homeland, you can be calm, strong and true stands the watch on the Rhine. He had done his part to sell war bonds.

THE STUDY

HOW DID IT HAPPEN? THE DOCTOR WAS WHISPERING TO Jakob's wife in the doorway to his study. The people who brought him home said he had fainted, and Jakob hadn't troubled to correct them. What really happened was that after the program he had fallen down, and they hadn't noticed until they nearly stepped on him! Let him rest for now, the doctor advised Helene. The professor is still confused (he wasn't), but he'll be fit again. After all, for a man of sixty-nine . . .

The director of the physical institute looked concerned as he bent over Jakob. They had had strained relations in the past, at times communicating only by written messages delivered by the porter. But their relations had mellowed, and now the director badly needed his associate's services. Who, after all, was there to replace the institute's theoretical physicist in these short-handed times?

Despite Jakob's condition, the director insisted on amiable conversation. As he saw it, the war should leave Germany dominant on the continent. Didn't it now dominate the Ukraine, the Crimea, the Baltic provinces, Poland, Rumania, and Finland? With the breakup of Russia, didn't Germany control Eurasia from the Oder and Vistula to Vladivostok? With Ukrainian grain to feed the German people, Rumanian oil and iron from Lorraine to feed German industry, and a colonial empire in Central Africa, wasn't Germany a world power equal to Britain and America? Wasn't that the point? Of course, to reach their highest development, Germans must not spare themselves nor their leader spare them. Leaning close to his silent associate, the director asked: Would Germany's position assure the future of the monarchy and empire?

The old professor raised himself on an elbow to reply: The facts! Isn't the Geheimrath overlooking the facts? Don't we live in daily fear of a catastrophic breakthrough in the West now that America is sending unlimited divisions against us, Bulgaria is out of the war, Austria-Hungary is on the verge of collapse? Doesn't chaos threaten here at home? Hundreds of thousands of workers in Berlin have struck for food, peace without annexations, and democracy; the cities are starving; soldiers are deserting; the Emperor is defiled, and the monarchy is viewed as an obstacle to peace; the Crown Prince is a profligate, and his brothers are scarcely better. The Geheimrath misses the point. The future of Germany doesn't depend on influence in Turkey or a piece of Belgium. It depends on character, on spirit, on idealism—on, among other things, science!

He looked up at the Geheimrath to emphasize this last point and found himself looking at an empty chair. He is still very weak, he heard his wife say in the vestibule. He has difficulty focusing on conversation for any length of time. Jakob realized that the Geheimrath hadn't heard a word. But perhaps he hadn't said anything.

The more he thought about it, the more the old professor became convinced that his weakness had been brought on by preoccupation with the war. He went to bed with the war, got up with it, was kept awake half the night by it. He tried to penetrate to what was essential in these times, which made it all but impossible to carry on with his work.

The latest distraction that set the faculty to quarreling was the prorector's attempts to get them to agree on the wording of a statement of their unified will to work for the honor and freedom of Germany, which he would then cable to the Reich Chancellor and the General Field Marshal. In this fifth year of the war, the fiftieth month, no one had any fresh ideas any more, and the old ones couldn't be discussed with any hope of reconciliation.

Droning on about U-boats and about how they had taught people the meaning of technical progress, the director of the physical institute had started the row at the prorector's latest meeting. It was just hours before Jakob had become ill. He could still hear the shouting.

I can't believe that anyone still puts his trust in U-boats, someone had cut short the Geheimrath. Militarily, they have been disastrous, bringing America into the war, and, humanely speaking, they are the worst violation yet of civilized restraint. To sink ships on sight without warning and without means for saving survivors is worse than bombing undefended towns or using poison gas and exploding bullets.

Other members of the faculty tried to make themselves heard then. The prorector even thought it was a good time to get to the business at hand, but he didn't know the director of the physical institute as well as Jakob did. The Geheimrath shouted on cue that U-boats are humane, since they bring the end of the war closer, especially if the U-boat war is waged with all possible force. Besides, U-boats are going to make wars impossible in the future. They have already caused such panic that navies have all but disappeared from the seas.

So it had gone. Jakob turned to his wife who, he believed, had brought her needlework to his study to be near him: The madness of it, to believe that the world will be at peace when the technology of warfare is complete. She raised a finger to her lips. They were not alone. From the alcove he heard fragments of conversation—the cultural man . . . the man of force . . . failure of nerve . . . Moloch of democracy . . . revolution . . . Bolshevization . . . Americanization. That decided him to lapse into simulated half-consciousness.

After Helene had shown the visitors out, she returned to the study to find Jakob waiting for her, alert and with an apology for saddling her with the burden of entertaining colleagues. They repeat these slogans interminably, he said, but they won't analyze them. There's no talking to them!

Helene explained: They're younger than we are. She continued, seriously this time: Perhaps they're excited by the material and superficial aspects of the war. You're always telling me how busy the war keeps the Geheimrath, how important he has become, or thinks he has. How important *physics* has become! he corrected her. The Geheimrath has not suffered because of the war as Planck and some of your other colleagues have, and so he hasn't had to depend on his physics as they do. She said this as a question.

Poor devil, Planck. One son dead, the other a prisoner.

Finding strength he didn't think he had, Jakob began to tell his wife about the commemorative talk Planck gave in his capacity as rector of Berlin University two days after the beginning of the war.

It had been an occasion for recalling past glory and looking to future glory. Many students and faculty who came to hear him already had their marching orders to defend the homeland. Jakob had sensed the tension and excitement in the more than full auditorium as Planck moved to the lectern. Clearly, on this sensitive day the rector's proper task was to express the patriotic sentiment of the assembly. Planck, however, saw his task differently. Coolly, technically, he spoke about theoretical physics and philosophical questions it bore on. He didn't speak about the wars of liberation and the unification of Germany and only obliquely referred to the coming great war. His omissions were partially compensated for by the closing German anthem, which caused tears to flow liberally. But Planck did not escape criticism. As people poured out of the auditorium, Jakob overheard harsh words about Planck's failure to address the war.

They didn't want to be sent off to the front with a talk about physics, about . . . What did you say his subject was?

Dynamical and statistical laws. But also conscientiousness and loyalty. If I remember his point, in science and far beyond it, conscientiousness and loyalty are guaranties that the individual will achieve the highest good of the human intellect, which is inner peace and true freedom. Conscientiousness and loyalty, the guides for every German at the post to which fate has assigned him. It was Planck's duty to speak of physics on that occasion, as a *physicist.*

I should have talked of physics, he added after a moment. When? she asked. At the Patriotic Evening. I let myself be distracted by all the talk about war. I lost sight of . . . Your post in 1870 was on the battlefield, she reminded him. You spoke of doing your duty then. That was appropriate. He shook his head: My whole life since then has been in physics, and that is what I should have talked of.

As if she were afraid that excitement was taking hold of him, she rose just as he was about to get up from the couch (he never gave any audience an exposition of his thought except on his feet). She suggested a brief rest before tea. As she left the study, he could see that she had little hope of having successfully interrupted his train of thought.

The old professor had a great desire to talk about his life in physics. His life in *classical* physics, as he liked to describe it, though he half took back the word every time he said it because it flattered rigidities of mind he wasn't proud of. His colleagues spoke of Kirchhoff, Helmholtz, Hertz, and other predecessors of his as classicalists and meant that their thought belonged to classical physics. He found that way of speaking acceptable, and yet he was troubled by its suggestion that people knew precisely what classical physics was. In the time of the classical physicists, the question of what it was had been one of the liveliest questions. Apart from universal principles and absolute constants—those rare and precious supports of physicists' faith that they discover truth about nature—physicists hadn't agreed about much during the fifty years Jakob had followed the progress of physics. No, physics had not been any one set of theories, principles, concepts, problems, or methods of research. As today, physics in the past was no one thing.

His life in classical physics.

Most of it he had spent right here in this or another

study. Its relics were all about him, the lectures and papers he had written while standing at his desk near the window. He remembered how at this same desk he would light a gas lamp or—he could remember that far back—even candles in the late afternoon. Now electric lights would dispel the gloom in his study as soon as his wife brought him tea.

And how he wrote during those years of hope! Then, as not always now, he completely rewrote his lectures each year to keep abreast with discovery. Sometimes it was past midnight when he had finished writing out the next day's lecture, and even then he would not stop but would pull from the drawer beneath the desk top the manuscript he was working on and would begin again a lengthy calculation that he had got nowhere with the day before.

The desk smelled of wax. The blotter had been replaced by a fresh one, and his pens were lined up pointing the wrong way, their tips to the right.

He lowered himself in the stuffed leather chair, where he had often had his best ideas in physics. Nowadays he was more likely to daydream there or even sleep. He was glad for the momentary release from his feelings about his worth as a man and as a physicist (they came to much the same thing). Cradling his head in the worn hollow of the chair and with eyes closed, half-drowsing, he recovered an image of tranquility. He saw himself as others did, as a German professor, a theoretical thinker, freely and productively occupied with research and teaching. If what he saw was a picture from an age that was passing, it was nonetheless a noble picture. Hadn't Kirchhoff been a thinker like himself who lived his life in the quiet setting of science, in the lecture halls and laboratories of German universities, far from the turbulence of social and political action? Hadn't Carl Neumann been another, a professor for whom work was everything, for whom teaching and research excluded even the world of university affairs and assured him public anonymity? Or Wiechert, his life passing in simple forms, a good marriage and sufficient well-being to assure him the quiet and security for the scientific work that expressed his total personality?

In this way, one after another colleague presented himself to Jakob, until they all faded before an image that to him summed up all their lives. Meeting on a walk in Heidelberg one day, Kirchhoff asked Helmholtz if he had noticed the peculiar light reflected from a rough sea at sunset, and then for half an hour the two physicists stood thinking about that while Helmholtz' wife stood thinking about those peculiar creatures, physicists, all three standing for half an hour in soaking rain.

With his colleagues Jakob had shared the life of the German professor that made such mental concentration possible. But nowadays, the distractions!

When Helene brought him tea, he was looking at the gallery of physicists' pictures in his study, which she had arranged in a pleasing pattern. They were reassuring to him, especially Helmholtz' picture, which was a reproduction of the painting Lenbach made soon after Helmholtz had come to Berlin to head the physical institute.

She looked at Helmholtz, too, at the intelligent eyes, the broad forehead, the majestic calm. Jakob had often pointed out these features of Helmholtz' intellectual greatness. Just as I remember him, he said. The Reich Chancellor of science, Lenbach called him. And Jakob's great teacher.

Directly above Helmholtz' portrait was a photograph of Maxwell, whom he had never met, to his regret. Below and to the right was a photograph of Helmholtz' great pupil Hertz, looking fresh, youthful, confident of his powers.

Beside Hertz's was a photograph of Drude. Jakob looked away. He blew on the steaming tea his wife brought him and studied its surface.

To the right of Helmholtz was a picture of Kundt, Jakob's other great teacher. And opposite it was one of Kirchhoff, whom he had admired from a distance. He had sent him one of his early mathematical physics papers for criticism—and praise. Kirchhoff wrote back that it would be pleasant to be kind, but that it would not be honest. What he said was: Hurried too much. Oversight. Poorly chosen expression. Wrong! His words still smarted, especially as Jakob recognized their truth.

Several times Jakob also thought of sending his work to Hertz. But there was always a new difficulty just when he thought he had it right, and he put off sending it. Then he heard that Hertz was ill, then that he was dead. At thirty-six Hertz was dead. And immortal! Several years older than Hertz, Jakob for the first time confronted his own professional unfulfillment and, frankly, his mortality. He also sensed that, with Hertz's death, German physics had lost its next leader and physical research its most promising direction. When Helmholtz and Kundt died that same year, Jakob wondered where German physics would go next. In his forties, he worked as stubbornly as ever, but he knew that others such as Planck would be the next leaders.

Jakob was still thinking of Planck when he awoke the next morning. Planck had called on him and on all scientists to persevere at their scientific posts in the midst of the temporary upheavals. Jakob knew his post.

He felt almost his old self again, which may be why he decided to give a talk—at his post. It would be a talk to the university, which meant that he had to see the prorector for arrangements. He would see him this morning, only without telling his wife, who would only appeal to his reason and urge him to wait until his doctor had come. Reason would drain away enthusiasm this morning.

As he dressed to go out, he thought about the talk. He must begin with the most important developments in physics in his lifetime. The recent move away from mechanical representations of nature was a good place to start, though he knew he would be misunderstood. A good half of his audience—historians, philologists, and the like—would be illiterate in natural science and would take him to be talking about the overthrow of materialism, which would give them a warm feeling. There was not much he could do for them. Jakob disapproved of materialism as much as any student of Kant and admirer of Helmholtz and Hertz, but he didn't approve of the denigration of mechanical physics through its confusion with a philosophy of matter and force. After all, he had spent half, the better half, of his life working within a theoretical physics integrated largely through mechanics.

As Jakob reached for his coat and hat on the hall wardrobe stand, his eye fell on a walking stick hidden under his loden cloak. He pulled it out with the idea of using it, although he had never done so in town before. He noticed the inscription on the metal band, and for a moment he felt that Drude was standing there beside him. He set the walking stick down again gently and took only his coat and hat.

When he reached into his hat for his gloves, he found only one. He searched in his coat pockets, on the floor, then in his study where he might have dropped the other glove on the evening he was brought home sick. An irregularity of this kind in his daily routine could only have an exceptional cause. The war. Inattentive servants worried about relatives at the front.

REASONS FOR PERCEPTIONS OF THE INADEQUACY OF THE MECHANICAL WORLD PICTURE, he had written at the top of the page. It caught his eye in his search for the glove. He took up the pen. Repeatedly, he added, great theoretical advances in physics arose from attempts to interpret the world by the principles of matter and motion.

Twenty years ago he was still making such attempts himself, but by then they were usually seen as antiquarian. He suspected that the satisfaction he got from working within the older mechanical physics was rooted in metaphysical prejudice, which was hidden from him. In any event, when he looked back it was hard for him not to see the move away from mechanical interpretations as the beginning of the move away from classical theoretical physics itself. Everyone conceded the unparalleled clarity and logical rigor of mechanical constructions, qualities which had much to do with what Jakob understood by the word *classical.*

But Jakob's immediate purpose was not to resurrect an inadequate physics. It was to set down scientific reasons that had been given for the failure of mechanical explanations in certain parts of physics, and that was no small assignment. The mechanical world picture had too many resources for merely one or two objections to undermine physicists' confidence in it. Jakob rapidly wrote out several objections without regard to their weight or sequence.

The mechanical world picture was incomplete, as even its advocates had conceded. The extension of mechanical principles and concepts to theories of heat and electricity led to disagreements with experience. Mechanical explanations were objected to in principle, since if there was one mechanical representation of a phenomenon, there were always others. In fact, there were so many reasons why physicists had grown disillusioned with mechanical explanations that Jakob decided to write down only one or two more and be done. There was, for instance, the difficulty with analogy. The new understanding of mechanical explanation (or description) was that it was an analogy between a mechanical system and another physical system, and analogy was a two-way street that allowed explanation *of* mechanical systems as well as *by* them. Jakob didn't actually write down this last difficulty, since he sensed that the discussion was already becoming too abstract for his general audience. And he hadn't even come to objections to mechanical explanations from the point of view of his own special concern, the world-ether, the seat of electromagnetic waves. These objections were of long standing and formidable, but he felt that to list them all would be querulous in view of his sympathies and his conviction that the greatest physicists—Helmholtz and Hertz, to mention only two—had all looked to mechanics as the foundational discipline.

He heard the clock strike ten. He was still standing in his coat at his writing desk. Too late, he remembered that he had meant to go out. The prorector would be at his second breakfast, and the old professor expected his doctor before long. The enthusiasm he felt earlier had left him.

He picked up a textbook on theoretical physics he had just received in the mail and read that until recently a physical explanation meant simply the reduction of what was new to what was familiar according to the tenets of the mechanical world picture. But the diametrically opposite program of subsuming everything under electromagnetism—he was again writing notes for his talk as he read—that is, the program of the electromagnetic world picture, had arisen and gained strong support, and in the present state of science there were no grounds other than historical and pedagogic to continue to place mechanics at the head of theoretical physics. That was well put and would go at the end of this part of the talk.

When his wife came in, he was lying on the couch, still in his coat. Before she could remark on it, he told her about the missing glove. But she was busy pushing the papers on his desk into a neat pile. The porter from the physical institute was there and insisted on seeing the professor himself. What was more irregular, he had brought a cat.

The porter had brought Maxwell. The stray cat had befriended the old professor one morning on the way to the institute, followed him inside, and settled on his desk, where he apparently meant to stay. As the cat liked to scratch his back against Maxwell's weighty *Treatise on Electricity and Magnetism*, Jakob took to calling him Maxwell. That was part of the problem. It turned out that even a cat could cause disagreement between theorists and experimentalists. For a time, a party at the institute, led by the director, called the cat Röntgen. After all, it was a German cat, whose nose for discovery took it to every corner of the institute. In the end, the cat decided for the old professor (who fed it), and his name came to stick. Maxwell was still unpopular with the opposition in the institute.

Now on the floor of the study, Maxwell was as quiet and thoughtful as any theoretical physicist. The porter's somewhat confused account of his reasons for bringing the cat to the professor's home had to do with the custodian of the institute, who was known to detest the cat and even to kick it on occasion. This morning the opposition, acting through the custodian, had come down on Maxwell's head—and nearly smashed it. Fortunately, they had only smashed his spirit, and only temporarily at that. The porter had done well to bring Maxwell here while his defender had to be away from the institute.

When the doctor arrived, he noted that Jakob's strength was returning, and with that positive medical judgment to back him up, Jakob told his wife about his work on the new talk. What was so troubling to him about her response was not that she opposed the idea—she didn't—but rather that she reacted as if he had said something sad.

The food they got these days! The scraps! He complained about them as he followed his wife into the dining room. From morning to night, his stomach didn't leave him alone, and even physics couldn't distract him for long. His weight was down from one hundred and eighty pounds to one hundred and twenty, and his doctor told him that his diet caused his heart ailment. The delicate letter-scales on the dining room table were sturdy enough to weigh out the scanty daily rations.

He picked up the newspaper on the table. Instead of reading news of the war and political editorials or music and theater reviews in the feuilleton, he looked for notices about food. Berlin had prohibited the sale of sauerkraut, embittering the Bavarians. Well, what did they expect in Berlin? Milk was going up in price, which didn't matter much, since all he got was an eighth of a liter. An ad for dried apple cores and peelings, one mark per kilo, suggested to him a tolerable soup. He read that there would be plenty of onions at least.

Their meal was without meat again, another official meatless week. In place of the regular weekly half pound of meat there was a little extra bread, looking grayer than ever. He broke off a piece and studied its uncertain ingredients, then spread a little turnip jam on it. He scooped up his ration of watery war soup and then with resignation chewed through the pound of potatoes on his plate. It was over all too soon. He saved his milk for Maxwell.

It will be onions for us next, Jakob said to Maxwell as he put the saucer of milk on the floor of his study. You and I are coming apart just when we need our strength most.

Even the preparations for the talk on physics proved to be unsettling, for they inevitably led him to think of his own intellectual and professional development. Intellectual development? Yes, he could say that. Professional development? In the morning mail, a questionnaire from Poggendorff's biographical dictionary asked him about that. Rubbing salt into wounds.

Jakob had moved from the mechanical to the electromagnetic standpoint in physics sometime after Hertz's work on Maxwell's theory. The photograph of Hertz on the wall over there—he had asked Hertz for it at the time he had asked him for advice on how to produce electric waves in air with prisms and lenses. Soon after that, he saw Hertz at the Heidelberg meeting of the German Association of Natural Scientists and Physicians, where despite his tender years Hertz mingled with Helmholtz and the established physicists when he wasn't surrounded by younger admirers. Naturally Jakob couldn't get close to Hertz at this time, but the meeting had been a turning point for him all the same. Hertz's great address there on the unity of the physics of the world-ether inspired him.

At first Jakob tried to derive the equations of the world-ether from mechanical first principles, and he more or less succeeded, but then so did Boltzmann and others, and he didn't bother to publish this work. Then for a time he followed Hertz's example of beginning with the bare differential equations that described the behavior of the world-ether. But in the long run, this mathematical phenomenology could not satisfy him any more than it could Hertz.

In general, he remembered those years before the turn of the century as bewildering for physicists. There were so many ways physicists could go. At one and the same time, Jakob might be studying Helmholtz' derivation of the basic differential equations of physics from a common least-action principle; Hertz's reformulation of the principles of mechanics to allow all physical phenomena to be traced to them; Boltzmann's universal atomism; Mach's phenomenological and Ostwald's energetic foundations of physical science along with their denials of atoms, forces, and mechanical explanations as the proper approach for understanding the physical world. There was much talk then about the new theoretical physics, which excited Jakob. Temporarily swept along by the vogue of viewing energy as the sole reality, he was soon brought to his senses by Planck and Boltzmann, whose writings showed him that the energeticists were hopelessly confused on physics and that they threatened to mislead young physicists into believing in a royal road to discovery.

At about this time, Jakob settled into his career. He decided to do what interested him most, which was to study the relations of the world-ether. This active medium pervaded the universe, unifying it, bearing all forces and yielding atoms as regions of abnormally intense states. It was a worthy theme.

Gushing, eternally bright/Flowing through all worlds.

The world-ether! That classicist of the German language could sometimes speak to physicists! It would go into Jakob's talk.

What finally decided him to derive the phenomenal world from the behavior of the world-ether was the electron theory, which was built upon Maxwell's theory and the now empirically confirmed atom of electricity and which went deeper into the connections of nature than the mechanical and phenomenological viewpoints had. The trouble was that the electromagnetic world picture soon ran into difficulties just as the mechanical one had, though Jakob did not think they were insuperable.

On balance, his work in physics left him with pleasurable memories, not just with regrets. He had taught and done research in physics alongside the best. But his professional life? How could the editors of the biographical dictionary hope to encompass any professional life in these categories and small spaces? Only a statistician could believe in these bare bones of a life in physics.

Family name:
First name:
Scholarly titles:
Your address?
Date and place of birth:
Relatives of scientific interest:
Earned doctorate when and where?
Studied when and where?
Former positions, giving date of installation:
Present position, giving date of installation:
Do you edit a journal?
Instruments invented or noteworthy achievements, giving year:

Titles of books:

Titles of articles on subjects from the exact sciences:

To answer the questionnaire was not a pleasant task for Jakob.

His family name Jakob caused people to assume that he was an Israelite by descent, and after they learned that he was a theoretical physicist they had no doubt. People had already begun to think of theoretical physics as a Jewish speciality, which it might become if Jews continued to be passed over for the important experimental physics jobs. If Jakob were to believe some of his colleagues, what kept Jews from advancing was the abstract-conceptual nature of the Semite. Or it was the cheek, the forward personality typical of Jews, if Jakob were to listen to others (not friends of his) who fell back on cruder reasons. Physicists at times sounded as if they were protecting their field from the plague when they did not want a Jewish physicist on the faculty. They described his work as unhealthy because his theories did not lend themselves to constructions and images, and they said his seductive Talmudic logic made him a dangerous teacher, a high priest who nipped in the bud any undeveloped ideas. What was wrong with having more men of critical intelligence in physics, Jakob wondered. There were always other physicists who could give life to their conceptual skeletons. Of course, when a faculty wanted a good young physicist who happened to be Jewish, they used the opposite argument with the ministry: since the man would be teaching objective physics, his Jewishness would not affect his subject at all. But it was always a wrangle when a Jew turned up in a favored place on a list of candidates. A suspicious name such as Jakob was enough to set in motion an inquiry by the ministry into the background of the candidate, which meant that, to start with, Helmholtz or Planck or someone else would be asked about the man's descent. No one admitted publicly to anti-Semitism as a principle, but Jakob had heard his share of confidences.

Where Jakob had grown up, there were few Jews. If a Jew had the misfortune to show himself on the street, he was soon followed by a mob. It was much the same if he were a Catholic. Only after entering the academic world had he got to know some Jews. Fortunately, the academic world did not behave like the mob, but still . . .

First name: Victor. His parents had considered heroic names for him like Hermann, regal names like Friedrich or Ludwig, liberal names like Robert. They settled on Victor, which sounded at once vaguely heroic, regal, liberal—but, unfortunately, English, which was a liability in an anglophobic time.

Mechanically he filled in the next three blanks on the questionnaire. He was born only a few months after the revolutions of 1848. That he should have come so close had always seemed unreal, for revolution was the opposite of all he valued in day-to-day life. Order, stability. There was going to be another revolution any day, they said. Would civilization never reach a level where people could live out their lives without having to experience chaos?

Relatives of scientific interest: None. After generations of preachers and minor officials, Jakob was the first scientist in the family. By contrast, many of his colleagues appeared in Poggendorff's dictionary surrounded by their forebears, descendants, and in-laws, which encouraged readers in their belief in the hereditary theory of scientific genius. Their names spoke for themselves: Eilhard Mitscherlich, Gustav Wiedemann, Eilhard Wiedemann. To sort out the nest of Hagens, Bessels, and Neumanns required the services of a genealogist. Jakob thought of the Kohlrauschs, the Helmholtzes, the Wiens, the Webers, the Königsbergers, the Sohnckes . . .

Jakob had opportunities—and had passed them up—to marry into scientific families. As a young physicist he was often invited to homes of professors of physics, chemistry, and mineralogy, whose daughters were the right age. He considered an early marriage a time or two, but his salary was so small it seemed imprudent. In time his opportunities diminished, and with them his interest. All the same he realized his liability, since reports to ministries of education on possible candidates for professorships sometimes drew attention to their single status, which in a man after a certain age was regarded as a serious eccentricity. When Jakob finally married, it was too late to make a difference to his career, though it did lift some doubts about his character. (Helene refused to this day to accept as evidence of previous doubts the sharp increase in the numbers of invitations from colleagues' wives after his marriage.)

In a sense Jakob did belong to a family of physicists. Although he had no children of his own, he had what he might call a scientific grandson. A young physicist—a former student and private assistant of his—whose budding career he had watched with great pride and hope only to see it destroyed by the war, had asked him to be godfather to his child. Perhaps one day the child would carry on for his father—and for Jakob.

Questions about his studies. The last time the dictionary had appeared, its editors had chided the scientists for extending the When? of their studies back to their childhood. How typically German to give such weight to individual development! Each his own Wilhelm Meister. To Jakob the important question was not When and Where? but With Whom? When and Where could no longer identify the decisive influence on a scientist, which was the individual professor who guided him. Teaching staffs had grown too large.

Helene brought him a telegram from the ministry of culture, which had been held up for a day at the institute and even opened there. She was angry at the disregard shown her husband. To add insult to injury, the director's apologies were conveyed by the porter.

Helene didn't understand. At one time a telegram from the minister's officials would have been the signal for negotiations for a better job. Now it was merely a routine expression of official concern for his health. Jakob knew that much without looking at the telegram, and the director had known it when he opened it. It was not a message to be handled with care. Nevertheless, Helene was right.

Present position: Honorary ordinary professor for theoretical physics. His position had changed since the last questionnaire twenty years ago, but it was a change in name only. An honorific title. When he was appointed extraordinary professor for theoretical physics in a newly created position at a small university, he never dreamed that it would prove a dead end for him. His failure to advance at his or another university would not be hard to explain if he had been an angry Czech or Pan-Slavist—in which cases he wouldn't have got a job in the first place—or if he had been a Catholic in the north or a Jew anywhere. (When he was young, he believed that if his name had been Schmidt or Müller or Meyer he might have become director of a physical institute. Jakob trusted that the ministry's discreet researches had determined that his ancestors were Christian for several generations back.) He believed that he had mainly himself to blame for a lackluster career. But he also had his job to blame, in part, since extraordinary professorships for theoretical physics had proven dead ends for more than one deserving physicist. Ordinary professorships and institutes for theoretical physics were rare, and most physicists who advanced did so by way of experimental physics and outstanding published research.

While a student, he had read in one of Helmholtz' essays that life acquired meaning from action alone, not from the passive accumulation of knowledge from books. The remark encouraged Jakob to enter the world of action by sending off a portion of his doctoral dissertation to the *Annalen der Physik*, where it was promptly printed. Earning a small keep from fees by lecturing as a Privatdocent and by working as a lowly salaried assistant in the physical institute, he applied himself unsparingly to research. He published several papers containing experimental results, one or two of which he later learned were judged good for a beginner. The other ones were judged moderately interesting and evidence of conscientious application. In sum, his early publications didn't catch the eye of chair holders, but they were sufficient to earn him an unremarkable junior appointment in theoretical physics. His later publications did not substantially revise the earlier judgments.

No one is dishonored if he does not become an eminent discoverer, the professor consoled himself. Inner gifts set apart certain physicists, whose unstinting efforts lead them to a deeper understanding of nature. Poets have said that noble personalities lie in their being and not in their accomplishments, and if that is so then the worth of a physicist cannot be measured by his published research alone.

His scientific spirit, love of truth, and dedication might inspire his students to surpass him.

Was that what he wanted his obituary to say? Had everything already been said? Was there to be nothing more?

The questionnaire was to blame, of course, for these thoughts. It asked for the same facts as obituaries, only without the same sympathy. There was even a footnote in the questionnaire asking for the date of death *if* the scientist had died since the last edition of the dictionary. The whole thing was irritating enough—though, looking at the dull, official form, he couldn't quite say why—for him to consider filling in the date of his death. Today? Tomorrow? Next week?

Next question. Do you edit a journal? No, never, he added for emphasis. He attributed a measure of the collegial feeling he retained in old age to having never edited a physics journal. He knew that authors and readers often overvalued their own work and undervalued others', which meant that editors worked within a controversial milieu of exaggerated hopes, fears, and vanities. He was not asked to edit, but for that thankless if indispensable labor it was almost enough if one were willing to take it on. (That wasn't quite true, he knew. He felt sorry for his colleague Eilhard Wiedemann, who had expected to inherit the editorship of the *Annalen der Physik* from his father. It passed to other hands, which Jakob believed contributed to Eilhard's complicated feelings about—indeed, his virtual abandonment of—physics.)

Of all the editors Jakob had known, only Poggendorff was born to the task. Poggendorff collected, catalogued by author and title, and bound in hundreds of volumes all of the dissertations, articles, and notices that passed through his hands. He had an elaborate index for the thousands of authors entering the biographical dictionary that Jakob was responding to at this moment. His orderly nature found its complete expression in the *Annalen der Physik*, which he edited for over fifty years, driving his competitors from the field. With the serenity and authority of one whose being was filled with a sense of order, Poggendorff guided German physics through its most consequential development. Jakob reflected that, in an earlier day, such an individual embodied all the organization that physics needed. Today the trend was toward external organization, and physics journals were not spared. The decisive, if fallible, editor was bound to be replaced by faceless editorial boards of refereeing specialists, which was all the more reason for Jakob to be thankful he was not an editor. He had been there with Kirchhoff and other physicists to see old Poggendorff into the earth.

He answered the rest of the questions quickly, since he had no new work to list.

Invented instruments, noteworthy achievements (meaning the discovery of a new planet and the like) . . . He didn't mind his negative answers here, but he regretted having to leave blank the question about books. If only he had acted more quickly! At the beginning of his career, he had recognized a trend toward the creation of positions for theoretical physics. With good reason he expected that with the growing separation of teaching into theoretical and experimental parts there would be a need for textbooks that treated theoretical physics comprehensively. Since there were no such textbooks then, he decided to publish his own lectures on mechanics, acoustics, heat, electricity, and the other parts of physics. Of course, he would do it only after his lectures had fully ripened. He was not deterred from his plan when in the 1880s some of Franz Neumann's pious students published his lectures on theoretical physics, which by then were mainly of historical interest. In the early 1890s an edition of Kirchhoff's lectures on theoretical physics came out, and although they had much the spirit of his teacher Neumann's lectures, they were more up to date. Jakob felt mildly discouraged, but since Kirchhoff's lectures did not yet recognize the full import of Maxwell's electromagnetic theory for physics, he believed there was still a place for his own lectures. His completely rewritten lectures were sitting on his desk when he learned that Helmholtz' recent lectures on theoretical physics were to be brought out by former students in a handsome, multivolume edition. It was Helmholtz who had been instrumental in introducing Maxwell's theory into Germany, and so his lectures would not fail to treat this topic amply. As the published lectures began to appear in the late 1890s, Jakob knew that they were the most authoritative presentation of the major results of nineteenth-century physics that physicists were likely to get. *Classical* physics in all its purity and lucidity! (He used the now fashionable word that relegated to obscurity at the same time that it praised.) Jakob had no intention of following Helmholtz' lectures with his own, for he didn't begin to have Helmholtz' grasp of theoretical physics in all of its connectedness. Helmholtz' lectures still left room for concise introductions to theoretical physics, which Voigt's textbooks admirably filled. Jakob had waited too long. He had no books to enter in this space in the questionnaire.

Jakob passed over the question about articles, because it referred to articles published in obscure journals that were not regularly indexed by the editors of the dictionary. With one or two exceptions, everything he had published appeared in the *Annalen der Physik.*

The current volume of Poggendorff's biographical dictionary was lying on his desk, open to his own entry. The modest three-inch column containing his vita and abbreviated titles of his publications was respectable, but only barely. The entry in the coming volume would be briefer, since besides his vita it would list only those few articles he had published in the intervening years. He looked at the row of Poggendorff volumes on his shelves, where nearly every physicist ended up in his alphabetical place for all the world to judge. Needless to say, most physicists appearing in these volumes had not produced much, and yet their lives in physics had surely begun as hopefully as his own. Now, with one-inch or two-inch—or even three-inch!—entries, the unfulfillment of the dead (and the dying) was final. He was looking at a graveyard.

Jakob decided not to return the questionnaire. He didn't want to be judged by his occasional published researches into side effects. For what really mattered, the questionnaire left no space: he had never let the fundamental questions out of his sight.

He pushed aside the questionnaire with its implicit reproaches. However the world might judge him as a researcher, he didn't doubt that he had devoted his life to the most fundamental—and most demanding—discipline. He couldn't imagine another life for himself, certainly not in business, concerned as businessmen were with the most arbitrary value of all, money. Nor could he imagine himself a lawyer, a pastor, or a physician, whose work centered on human needs. Nor an artist, nor a poet, who were captive to taste, a contrivance of culture that lacked the authority of nature. Music was different, for it evoked the harmonies of nature. To imagine himself a musician was not impossible, but musicians were born and he was not born one.

He was tired and yet knew he wouldn't be able to sleep. As he had often done before, he sat by his study window and looked out over the sleeping town.

He thought: Death wasn't content with victims on the battlefields, but had taken to the air and was even reaching behind the front to sleeping towns.

He remembered another town with completely dark streets, protected. It wasn't asleep. On the contrary, he heard foreign voices all around him as he felt his way along to the movie house. He, a German professor, had come to give them a view of the promised land of science. The attentive faces of the hundreds of academically trained soldiers before him—raised from the trenches—sank into darkness as the house lights were dimmed. On the screen a glowing image was projected, a shimmering sphere resembling an orange. Jakob raised the pointer toward the image and explained: You see before you the ideal of a closed circle in physics, of a total understanding of the physical world. It was unbearably alluring. And forever unattainable. Oh, yes, and the pointer was Drude's walking stick.

Jakob heard his wife say good night to the housekeeper in the hall. Then she opened the door to the study, but seeing the room dark she apparently assumed he was already asleep and went away again. She was used to his need of solitude for his concentrated work and had developed her own quiet daily routine, which she willingly broke only when he needed her. At other times, he would have followed her into her little drawing room, where she spent the late evening in warm light, reading or writing notes to friends or visiting with him if he felt like it. Tonight his mind was on his talk on physics, and he didn't feel like sleep. He switched on the light and returned to his desk.

Since he had emphasized disunity in physics up to now, he turned to unity, heading a new page INTEGRATIVE THEMES. After all, physics held together, despite its recent tumultuous development. Proceeding as topics occurred to him, he began a list of things that belonged to many parts of physics: experimental means (vacuum pumps, etc.), theoretical methods (ideal processes, etc.), mathematics (potential theory, etc.), principles (least action, etc.) . . . concepts. Here he began a longer list.

CONCEPTS

Inertia
Mass
Acceleration
Charge

Did the concept of energy promise a unified view of nature that was unattainable through any other concept?

Or: the world-ether. Since light, radiant heat, and electromagnetic phenomena all arose from a common source, was the world-ether the unifying concept for all of physics? Were forces and atoms reducible to actions in a world-ether?

Electron. Was the electron the Ur-atom of all matter, so that electricity became the foundation for mechanics, gravitation, molecular theory, and more?

Concepts of modern physics. The quantum of action penetrated and in a sense united many parts of physics, and if the latest works were to be believed, it had a fundamental connection with the relativity principle: every change in nature corresponded to a definite number of quanta, independent of the observer's reference system.

Since the total physical world was the subject of theoretical physics, a total theory was its ideal. Physicists spoke of world-picture, world-ether, world-function, world-postulate, world-point, world-line, world-this-and-that. As he wrote down these lofty ideas, he reminded himself that he must not give his audience the wrong idea of the everyday work of a theoretical physicist, which usually had little to do directly with the unity of physical knowledge. If the physicist flattered himself that his science supported the broadest generalizations known to the natural sciences, that was because it connected physical facts through a small number of exact, fundamental concepts and laws.

Jakob reached for a cup of cold tea, forgotten on the table to his left. As he lifted it, he lost his grip and spilled the tea. As he moved to protect the old oriental rug in front of the desk, he slipped and almost fell. Shaken, he started looking for a cloth to wipe up the spilled tea. He rattled the door of the cabinet before he remembered that it was locked. He picked up the cloth he used to wipe his pens but found it too inky, and he finally used his handkerchief. Kneeling on the floor and watching the fine white linen become mottled as it soaked up the tea, he knew he was adding foolishness to clumsiness. His left arm was still shaking.

Just before this accident, his thought had been to use lines from *Faust* at this point in his talk: Everything perishable/Is only a likeness. He used to be able to recite all of Part One of *Faust*, a common feat of memory for people of his generation but not for young people today, who didn't seem to care. He met a young professor at a technical university who had never even *read* the play! At the gymnasium Goethe had been his favorite author, and since then he had read Goethe's works beginning to end several times. To this day, he left physics books behind on vacation and took *Faust* or Goethe's writings on Italy and classical architecture—along with a Latin or Greek author, usually Sophocles, and a Baedecker.

His audience would know that Goethe had been no friend of the direction of physical thought away from the direct perception of nature toward mathematical abstractions—toward theoretical physics, to be precise, Jakob's own discipline. There was really no problem here, since Goethe was saying in his poetic way that physical theories are pictures or analogies. That agreed with contemporary scientific understanding, in particular with Hertz's understanding. Bringing to completion Maxwell's thoughts on the subject, Hertz had impressed on physicists that any theory is only a picture of nature, an analogy. Of course, Jakob must make clear to his audience that physicists' pictures of nature were not like pictures of the Rhine; physicists' pictures were physical theorems and concepts, which was something different. At the same time, physicists' pictures had beauty just as artists' did.

In Jakob's judgment, some artists' paintings had no more beauty than that spotted handkerchief on the rug. By beauty, then, he wasn't referring to the naive daubs of the Expressionists. (Their ideas were equally naive. They talked bombastically of unveiling the reality beneath the reality. Physicists had long known such talk was meaningless, which was the whole point of their notion of pictures of nature.) Still less was he referring to paintings in the so-called Youth Style, with their undisciplined images of natural fecundity. No, he had in mind paintings with the shimmering clarity of Menzel's *Balcony Room*, which hung in reproduction in his study. With morning light streaming through its white curtains, with its pleasing proportions, with its truth, Menzel's little bourgeois interior gave Jakob the pleasure that his own study with its high curtained windows and always freshly waxed surfaces never failed to give him. (Through Helmholtz he had met Menzel in Berlin and later again in Italy, where Menzel was sketching a ship. Each nail must be in its place, each scratch, Menzel had explained. It impressed Jakob, who knew physicists with less conscience.)

Put into words, Menzel's painting offered Jakob a way of introducing to the general audience the idea of a physicist's picture of nature. Just as the artist returns again and again to the canvas to retouch the curtains, the door, the light, the physicist is never fully satisfied with the picture he is presented with. He accepts that the picture will have to be retouched again and again to make it simpler, more coherent, more complete. It will come to represent ever more, but never all, of the physical landscape. That doesn't discourage the physicist, who realizes that a complete picture would foreclose the possibility of new experience. The idea Jakob needed to get across is that the physicist doesn't confuse his picture with reality any more than the artist does his. The picture is only a sign of something out there.

Were the new atomic pictures Expressionistic or Impressionistic? This question Jakob directed at Maxwell, the inert lump of a cat in the armchair. Jakob wondered if he had an unused talent for explaining technical ideas to nontechnical people. He would try out Menzel's interior on Helene.

Hoping she was still up, he started to go to her. He opened the study door, but couldn't pass through. For

standing in the hallway was a ministerial councillor, who was consulting his watch and train schedule. At this time of night! Jakob wondered as he bowed. Even odder, when he looked up, the councillor was gone. He closed the door, his hand shaking again. Jakob was certain that the man was a spy, here to report back to the ministry on how long he was likely to last.

For a moment he lacked bearings, but he soon recognized familiar faces in the waiting room of the Ministry of Culture. Planck had been there for hours and was going home to write to Althoff from there. Kayser left to go to Althoff's house to demand to see him. Wiener got up to follow him out and to send around his servant to that all-powerful and all-feared man. These were brave actors. Otherwise, the room remained filled with long-suffering professors. They were all bundled up and some were beating their arms to keep warm. In one corner, an argument was in progress. A professor defended Althoff as excellently informed on candidates for university jobs and more apt to back the right man for the job than was the faculty, which was a many-headed monster ruled by nepotism and dogma. Another protested that when negotiating for jobs one shouldn't believe anything Althoff promised in conversa-

tion, since he often didn't deliver. Angrily, another accused Althoff of contempt for professors who trusted him and of greatly lowering the moral tone of universities. Then the professors relapsed into silent misery again, while the fortunate among them slept on.

As requested, Jakob had arrived at nine, and now it was four. He was terribly hungry, since he hadn't gone away to eat for fear of missing his appointment. He was short of sleep, too, because of the long train ride to Berlin. The cold and immobility had put him in a stupor when Althoff came toward him at last, crooking his finger. In an interior office, Althoff told him to sit on the sofa. Did he smoke? Althoff went on writing, talking as he did, barely audibly, so that Jakob had to lean forward to catch the words. Warburg best, Cohn, Kayser . . . Cohn best, Warburg, Kayser . . . Kayser best, Cohn, Warburg. He had been told to smoke a cigar, Althoff reminded him sharply. Now Althoff laid down his pen and began to lecture him on personalities and positions, and Jakob wondered why he was taking so much trouble until it dawned on him that Althoff might be confusing him with Hertz. Referring to the discoveries that had brought him fame overnight, Althoff asked where he wanted to go next. He was not saying that Jakob would have a choice, and indeed it was unlikely that he would be called anywhere. But in the event . . . To begin with, he must put Giessen out of his mind. And if he didn't take the ministry's advice, he could forget about ever getting another job in Prussia. But he shouldn't be an anxious Peter. The threatening voice now became friendly. What did he think of Kiel? Or Königsberg, a nice quiet town? Or Breslau? Or Göttingen? Or Bonn? Before they were through, they had talked about half the universities in Prussia.

The ultimate recognition was Berlin, which he should think carefully about. Althoff explained soothingly that he should consider taxes and widow's pensions in Berlin and examination money. And he should consider the distances: to spend his vacation in beautiful countryside would mean a long and costly trip. And life in Berlin was not the quiet life of a small town. But Berlin was the center of all that moved, and that was what mattered. Berlin was a world-city!

Berlin would pull him away from his researches, he replied, from the quiet and harmony he needed. That was why Hertz declined Berlin. Althoff looked at him sternly. How arrogant of him to compare himself with Hertz! Did he think he was worth as much as Hertz? How much did he think he was worth? What was his price?

When will you decide about my case, he asked apprehensively. The decision has already been dictated. Will you tell me what it is? You will hear at the proper time and there is no use asking me now. Althoff gave Jakob a mocking look and lighted a cigar for him.

He really should have felt perfectly at ease here in his own small lecture hall in the physical institute. But something was wrong today, and he soon discovered what it was. Althoff was there, nearly concealed among the students. He was sitting quietly, waiting, which was unusual since his custom was to pass through the institute like a comet with a tail of dignitaries. It was fortunate for Jakob that he knew what Althoff liked to see, pretty electric lights, pretty photographs of Swiss landscapes, the Jungfrau and Interlaken, pretty experiments. He cranked the siren on the demonstration table, but the sound was not pretty.

The old professor began to lecture: Slowly through the centuries the human spirit has grasped the principles of sound, but it was only in the nineteenth century . . . Scratching chalk on the blackboard, he wrote some mathematics, which started off promisingly but came to no sensible point. At the beginning of the nineteenth century, he went on, Germany lay exhausted after its wars of liberation. Gauss and Goethe lived then, our greatest countrymen . . . Students were talking to one another in loud voices, evidently uninterested in remote history. He raised his voice: Fifty years later, Germany had to fight for its existence once again. After great sacrifice, it emerged victorious and unified. Helmholtz and Clausius tended the wounded in that war. I was there too. I was just your age, but of course you couldn't have known that . . . The noise was deafening now. When you reflect on Germany's struggles in the past, you must feel as I do about the cynical contempt for our ideal striving in science and art and for the Emperor's rule. You meet it not only in foreign propaganda, but also in the streets and press. The sheer ingratitude . . . No one was paying him the slightest notice.

Above the din, Althoff's voice was heard, commanding respect and silence: Have you trifled away your life like the other professors? Professor: I have devoted it to theoretical physics, which has created subtle concepts for making intelligible our experience of the natural world. At last, humanity's inborn curiosity has begun to be adequately . . . Althoff: We aren't here to talk about curiosity. Professor: Physics has created exacting methods of thought, offering students a mental discipline they can get nowhere else, instilling in them a hatred of fallacious reasoning. Althoff: We aren't here to talk about ideal learning either. Professor: People avoid quarrels and save money and time if they apply to their lives the critical, patient thought that physicists apply to their research. Lawyers, physicians, and officials who are practiced in the mental gymnastics of physics are less inclined to take those well-known shortcuts . . . Althoff cut him off for good this time: Your colleagues tell me that the whole social order is overturned by the benefits that physics bestows on industry. Gasoline engines! Cameras! Streetcars! Telephones! Wireless telegraphy! They tell me that the whole foundation of high culture is our rule over natural forces and that each discovery of physics extends this rule, that without physics Germany couldn't defend itself against enemies and compete in peacetime with Britain, France, and America for world markets in dynamos and steel. They can thank me and German industry that Göttingen University has institutes for applied mathematics, applied mechanics, applied electricity, and geophysics and has ten professors of physics and mathematics instead of its old five. Personally, I'm interested in medicine, but I have done something for physics.

Patiently, the old professor explained: Nature is the inexhaustible inspiration for physicists, not human wants. If Faraday had set out to invent a phonograph, he would have failed because he would not have discovered the physical principle the phonograph depends on. Of course, of course, Althoff said. Everyone knows about my plan for new state scientific institutes in Dahlem, where scientists would be able to carry out pure research unburdened by teaching duties. But my question is: How do you think you have come by your own luxurious institute? Physicists know that physical institutes have received many millions from governments and legislators, who are concerned with capital gains and not with curiosity and ideal education. They understand that the impact of physics on life is responsible for its rise. Professor: These are all external matters . . . Althoff interrupted him: Do people become physicists now because it pays or because they have cravings of the heart? The professor tried to answer, but Althoff went on: I am concerned that our science be the best, that Prussia be admired by all the world. Jakob, what have you done to help me? Professor: I have been teaching physics and doing research, time permitting. You can look me up in Poggendorff's biographical dictionary. No doubt I would have gone farther if I hadn't always had a dependent position in the physical institute, which forced me to the mathematical side of physics. I don't even have a strong talent for it. Althoff now explained that the ministry knows what's best for a professor and his science. In a statesmanlike manner, it assembles the opinions of all the specialists. Althoff went on for a while about the correctness of the ministry's decisions. Then he came back to himself: I alone can know every specialist's opinion of every specialist, if I want, so I alone can have an objective overview of the whole. When faculties don't appreciate this and insist on their own mind, they make a mess of it, I can tell you. I seldom let them get away with it, for their own good. When they get mad, I can take it.

Althoff became friendly: At one time I had my eye on you, Jakob. Your name turned up regularly on faculty lists of candidates. It was third or fourth from the top, to be sure, but it was there and it should have risen. The professor asked: Did my name have anything to do with it? Faculties are prejudiced, Althoff began. The professor: But the ministry . . . Althoff finished his sentence: Is unprejudiced. It appointed a Catholic at Strassburg so that all three religions would be represented among the historians there. And it did so in the teeth of the opposition by the faculty, who said that Catholics aren't free and so are unfit for objective scientific work, the life nerve of German universities, and so forth. The ministry stuck to its principles and recognized that the university is for teaching as well as research and that the religious complexion of its faculty should reflect the people it serves.

Professor: I heard that the ministry looked up my grandparents' birth certificates. Althoff: The ministry appoints Israelites, if that's what you're getting at, and the faculty complains about that too. As you know, we made Warburg director of the most important physical institute in the country. Professor: The faculty in Berlin wanted Warburg and the ministry didn't. Althoff: That was for factual reasons, not for the ones you are suggesting. You, Jakob, should be the last to suspect us of prejudice. Almost no other field has so high a proportion of Israelites as your own theoretical physics. Professor: What you did to Arons was pointless and illegal—and cruel. Althoff: Arons was dismissed because he was a Social Democrat, of course, not because he was an Israelite. Naturally, the Emperor couldn't tolerate a physicist in one of his universities who supported a party seeking the overthrow of the state by force. The ministry simply carried out the Emperor's wish.

The professor missed some of Althoff's words, since at the mention of dismissing a Social Democrat pandemonium broke out among the students: . . . you've no doubt heard that I'm a tyrant, that I rob faculties of their rights, that I'm coarse, callous, rude, intimidating, punishing, that I take pleasure in humiliating professors. You don't protest? I admit that I don't mind starting healthy rows, but everything I do is for my work. I personally conduct negotiations for every important appointment, which causes me infinite work. You could never believe the com-

binations I work out, always with the view to reducing collisions to a minimum. I spare no effort. I am selfless.

To the right and the left, people in the hall were expressing ill will toward candidates. They said that this candidate leans toward superficial philosophizing. His teaching is boring and weak. His students go away and don't come back. He thinks of teaching only from the angle of money, lacking the desired idealism. He is a difficult person. He is a man of sponge and chalk, clumsy with instruments. He lags behind a host of others and is unworthy even of the lowly appointment he has. He hasn't the respect of physicists, and no student today would recognize his name. Not only does he publish nothing and has ceased to count as a physicist, but his lecturing omits all that is alive in physics today. It is easy to document his ignorance of the promising direction of atomic physics . . . Gradually, the professor came to realize that they were talking about him.

But it was time to stop this general discussion, and Althoff signaled to a student. Upon removing a rubber mask, the student revealed that he was not a student after all, but a balding man in early middle age. Standing, he began to recite: During the winter semester of 1905 and 1906, I was in this lecture hall on assignment from the ministry. I will read to you from my report on Professor Jakob's conduct in class: He is neither short nor tall, a point I will return to. He has the expected roundness of a man of his age with sedentary habits. He doesn't have thick shiny hair, which makes some professors look ridiculous to students. He doesn't have notable deformities, though his nose is remarked on by students. He wears glasses sometimes, but he doesn't constantly push them up with his finger, a distracting mannerism professors have. On the day I observed him, he wore an unfashionably short coat, which made him look taller than he is. Otherwise, he showed little vanity in his appearance.

I was lecturing on theoretical physics, the professor called out. The informer said he was just getting to that: In clarity Professor Jakob is no Planck, in vivacity no Kundt, in spellbinding power no Warburg. Nor does he have Drude's warm concern for students. When he lectures he looks out the window most of the time. Students stamped their feet to confirm this behavior of their teacher.

He seems most comfortable with his back to the class, the informer continued. His handwriting grows tiny as he reaches the edge of the blackboard. The day I was there, he stared at the demonstration apparatus a long time and joked about the butterfingers of theoretical physicists. It is reported that he is inept at hydrodynamic demonstrations and his students come to lecture with umbrellas, galoshes, and raincoats. The professor interrupted him: You're confusing me with Meyer at Breslau. The informer added: His students wear earmuffs to his acoustics demonstrations. The professor didn't even try to answer this absurd charge and settled back to hear the worst to come. But the informer only said: Students I spoke with told me they had heard better physics lectures and they had heard worse. Since I haven't studied physics since the gymnasium, I can't judge the content of his lectures myself. Althoff interjected that he didn't pretend to know physics either.

That was well observed on the whole, the professor said. But I object to the criticism of my coat, and the report about my nose is only to be expected. With their feet the students were heard from again.

Althoff made it clear that he didn't want to hear any more of the professor's complaints and signaled to another student, who like the first rose and removed his mask:

My physics professor invited me to report on Professor Jakob . . .

There was tittering in the hall, so Althoff explained that it was the summer semester of 1906, when there was much moving between jobs. So he had to extend his sources of independent and reliable information on candidates. For this purpose he turned to his trusted associates, who were painstakingly conscientious in their assignments. Professors who cooperated with him could expect nice fees as chairmen of examination committees, he added, laughing.

Six students attended his class and more stayed away, the informer reported. As he didn't notice I was in the room, I didn't introduce myself. I noticed you, the professor said. The informer went on as though the professor had said nothing: His subject was Maxwell's theory, and he spoke well of Maxwell, Helmholtz, and Hertz, but of no one who worked after them. Maxwell had long been dead, and Helmholtz and Hertz had been dead for ten or fifteen years, so his lecture had a decidedly historical flavor. Needless to say, it failed to treat many subtle points that had entered physics since his sources. His students told me after class that he was always like that, behind times. The informer sat down and replaced his mask.

Consulting the notes he had taken, the professor remarked: As you say, I taught the principles. Physics was unsettled around 1906; much of what was new was unconfirmed, even flighty I would say. I taught what was solid, not what was yet to be discovered.

As the professor was speaking, there was commotion in the hall. Looking bored, students were leaving en masse. Only about a dozen masked officials remained behind. Pens in hand, they seated themselves in a circle like numbers on a clock-face. In their center, Althoff began to dic-

tate: To His Royal Highness! With regard to the elevation of Victor Jakob to a chair, the ministry has determined against that course. Our research has given us a total picture of the man.

Professor, protesting: I am ready to retire and don't want a chair, and anyway no faculty would want me.

Althoff brushed aside his objection: I don't listen to faculties! The ministry alone is in possession of all the facts. He began to dictate again: To the Finance Minister! Pursuant to the usual policy of refusing requests for raises without a call to another university, Professor Jakob is to go on receiving the same salary.

Professor: I haven't asked for a raise, and it's too late in any case.

We never close a file, Althoff explained. New facts turn up, even after death, and they go into this file. He held up a bulging gray folder with Jakob's name on it. Then he got down to work again, this time to dictate twelve letters at once by turning from official to official: Professor Jakob lacks originality, the root deficiency. His interests are narrow, and he fails to grasp the deeper lawful connections of nature. He claims to be a theorist but only picks holes in received theories or else wanders off into philosophy.

Professor: I lacked a laboratory and apparatus most of the time, so what do you expect? Naturally, I have done more than my share of criticism of theories, which is not such a bad thing.

Althoff, not listening: He lacks organizational powers . . . lacks a great scientific personality . . . shouldn't be entrusted with an institute . . . Althoff shuffled papers and whirled as he dictated: No one has taught physics that way for twenty years . . . an embarrassment to German physics . . . uncommonly dry and pedantic.

Yes, yes, there is probably some truth in all of that, the professor conceded.

Now he had a question for Althoff: Why didn't you divide the Berlin chair? Althoff looked surprised: We nearly did. Kohlrausch demanded it when we offered him the job, and I was in favor of it. But the faculty had its way in the end. We appointed Warburg, who handled the whole institute and supervised thirty doctoral students at once. Professor: The Berlin job would have finished Helmholtz if Siemens hadn't made a new job for him. Kundt was hurried to a premature death by it. Warburg was often at the end of his immense strength. Drude had immense strength too. If you had only divided that job, the tragedy might have . . . The job had nothing to do with that, Althoff said curtly: Drude took a serious fall in the mountains and wasn't the same after. Not that simple, the professor protested. Althoff wasn't listening, since he was getting his staff together to go off for their morning pint. Before he left the hall, he turned to the professor: Will you drink to this—professors are like whores who go with the one who pays the most! To his aides on the way out, he added: They all have their price, which isn't high.

THE INSTITUTE

HE WAS ALREADY DRESSED WHEN HIS WIFE CAME IN. SHE shook her head as he gave her reasons why he was fit to go to the institute this morning to give his lecture. Mainly he was concerned to prevent the Privatdocent from standing in for him as he had done when Jakob stayed away one semester with a heart condition. The Privatdocent had come to believe that *he* was the theoretical physicist at the institute. The matter was delicate since the Privatdocent was a former student of the institute director's. As Jakob gathered up his lecture notes, he glanced at Maxwell drooped over the window shelf. Clearly the cat was not up to resuming his place in the institute yet, and until that vicious custodian was gone he had better stay away in any case.

The new institute building was a good half hour's walk from his home. For years now, one or the other chair holder in physics at his university had pleaded for a new building, and whenever the Emperor was expected on tour Althoff ordered the director to draw up fresh plans. It was not until just before the outbreak of the world war that the government finally allocated money for the building. From Jakob's point of view it meant that he had his own small laboratory space—which came too late to make a difference in his career but was a satisfaction all the same. His theoretical direction in physics had been in part enforced, and as if to recover a right balance he had taken to spending as much time in his laboratory as he could. The old institute building had been more convenient to his apartment, but it also had been close to the center of town, which meant that those nervously delicate measurements that are the life of a physical institute were constantly menaced by passing trams and autos. The new institute building was set in the middle of a garden and bounded by a meadow behind and a private road forbidden to traffic in front. It was reassuring that the university owned the land around, for that meant that the institute would be spared nerve-racking noise.

He walked past a colony of little Italianate villas—perfectly boring, he thought—where some of his colleagues lived. The physical institute was now coming into view, set alongside the new geological, chemical, and physiological institutes. The cluster of institutes and their grounds resembled those eclectic palaces that guidebooks direct you to when you travel through this or that duchy. The Gothic spire, the Byzantine dome, the Renaissance stairway in front, the Baroque curlicues all over, and other historical references were all there in full view. Of all the buildings, Jakob liked the physical institute best, which was not surprising since he had had a modest say in its design. It was the sensible practice of the government to give the director a pretty free hand in designing the building, ensuring that it accurately expressed his mind and any subordinates' minds he chose to consult, which in this case included Jakob's. On approaching the physical institute, he could see his touch, the balanced proportions of the facade, intimating classical architectural harmony. Descending from a triangular gable in the center of the roof, lines of stones imbedded in the masonry walls suggested portico columns. The judiciously measured horizontals and verticals of the facade presented to Jakob's eye a satisfying set of geometrical relationships, as necessary to it as a combination of tones is to a musical chord. All physical institutes were not designed so gracefully, and it had been through persistence that he had prevented this one from becoming a neo-Gothic fantasy like Breslau's. Together with the director and the university's master builder, he had planned endlessly, but owing to foot dragging by the government the director died before the new institute was built.

That was in 1905, the first and only time he had headed the physical institute. After the director died, Jakob was appointed his replacement for a year. When at the end of the year he was complimented by the ministry, it naturally occurred to him that he might become the permanent director. As it happened, in that same year six or seven chairs of physics fell vacant, at Breslau, Giessen, Greifswald, and elsewhere. From several universities Jakob received confidential inquiries, which were encouraging but had no value in their own right. Only an official call to a chair led to advancement, and that was so whether one accepted the call or merely put it on record for later reference. When all the moving was over, Jakob was—he could scarcely believe it!—where he had begun, and a younger man was brought in from outside to take over the direction of the institute. He hadn't realized how high his hopes had been until it was over and he collapsed. He had never felt so depressed about his life in physics. His doctor wrote on his behalf for a leave, and he went to Italy, where among the orange and olive groves his nerves gradually quieted.

Looking back on it now, Jakob could see what he couldn't at the time. He had escaped a great complication in his life. To run an institute was like running a business, which was not in the line of his talents. Daily he would have been importuned by the students flocking through the doors of the institute along with the researchers, assistants, mechanics, custodians, and porters, all with their tasks to be assigned and their complaints to hear. There would have been some satisfaction too, but he didn't need it. For him teaching and research had proved enough. His teaching had been conscientious, and his research had been solid and—some of it—cited, even if it wasn't grand discovery.

When he returned from his emotional convalescence, the academic senate promptly recommended his elevation to honorary ordinary professor, which helped him over his disappointment even though he knew it was only consolation. From that moment on, he was resigned to the prospect that he would never have an institute of his own, either experimental or theoretical.

Jakob immersed himself again in local affairs. Since the new director had exacted as a condition of his move a firmer commitment from the government to build a new institute, Jakob discussed with him the plans for the institute that he had worked out with his predecessor. The new director had his own ideas, among them a Renaissance facade, which was much in vogue. But Jakob soon realized that he would have been just as happy with an institute that looked like a foundry or barracks, since his only real concern was the stability of the foundations of the laboratories. So Jakob had his way with the facade. Utility was their main concern, and here he and the new director agreed completely. They agreed, for example, on the need for a separate wing and entrance for the vast numbers of medical and other students who came to take the beginning experimental physics course. The separation kept indifferent onlookers from poking their heads into the laboratories and upsetting finely adjusted apparatus and serious research.

As Jakob looked at the separate wing of the institute, he thought how sadly unnecessary it had proved. The small numbers of students taking the experimental physics course these days hardly shook the building, and there was precious little independent research they could interrupt. He consoled himself with the thought that students would return some day.

Jakob was sorry he had never lectured in this experimental physics hall, since his own was less impressive from a technical standpoint. To shade the windows, raise and lower the enormous blackboards, and dim rows of electric lamps in the ceiling, one had only to operate a bank of switches on the wall. For any form of electricity or for gas, water, or compressed air, all one had to do was turn to the proper outlet at the demonstration table. A projection apparatus stood in the middle of the hall among the gracefully curved and sloping rows of seats. To Jakob the whole installation was living proof of the dependence of today's physics on the technology growing out of yesterday's physics.

Jakob now entered the institute and passed the porter's lodge. He didn't stop at this floor or go downstairs, where the custodian lived among the rooms for constant-temperature research and precision measurements. He went instead to his office, which meant that he had to climb to the next floor. There he was by no means at the top of the institute, for above him were the room for the Privatdocent, the now unoccupied room for the second assistant, rooms for advanced students, of which there were only two now, and rooms for beginning laboratory courses. There was yet another, unfinished floor with more unused rooms for advanced students, living quarters for the mechanic, and the entrance to the tower, to which Jakob sometimes retreated when the director was in a bad mood. Originally the institute was to contain living quarters for the director too, since it was understood that inspiration in physics was not ruled by the motion of the sun and that the director needed access to all his domains, day and night. Even though the planned quarters were nothing like Leipzig's palatial twelve-room suite and veranda, they were viewed as extravagant by parliamentary critics and had to be dropped. Jakob now passed by busts of Clausius, Helmholtz, Hertz, Maxwell, Faraday, and Fresnel, then by portraits of physicists of more local significance. He passed the room for the first assistant, the library, and the director's office and laboratory. He approached the small lecture hall for theoretical physics and the adjoining preparation room, where he assembled apparatus and practiced effects he would reproduce before his students. Before he reached the lecture hall, he came to his small laboratory and office. He turned and went inside, where he sat down to rest after the long walk.

As he listened to a lonely pair of footsteps in the corridor, he recalled the institute as it was before the world war when it was filled with dedicated researchers. The new director had been determined to make the institute a credit to himself, to physics, and to the university. The director's own research from the institute soon became a steady—if, as Jakob judged, at times a shallow—flow of publications. Partly because of the attention it brought the institute, the number of advanced students grew. Most of them came from abroad, since German students preferred big universities. Taking inspiration from his old teacher Warburg, the director drove his students as he drove himself and, at first, tried to drive Jakob. Anyone who gave less than his full measure was made to feel unwelcome. Woe to any doctoral candidate who spent mornings on the tennis court and then dropped by the institute in the afternoon to take a few readings before going off with friends to the tavern! Early every morning the director made the rounds of all of his students and, like a field commander, heard reports of their reconnaissance into unknown territory. Each student had to be completely on top of his subject, for the director knew—by author, title, place, and year—nearly every publication that bore on the work at hand, and the consequences of being unprepared were not pleasant to contemplate. By dedication and fear, the researchers were bound together. Since at any hour of day or night the director might appear by their benches with a question or suggestion, they ate supper at their posts and kept awake by tea made over a Bunsen burner.

That kind of institute life had all but disappeared in 1914. Nearly the whole body of advanced students, assistants, Privatdocenten, and miscellaneous staff were called up or volunteered or left the country. Although an unexpectedly large number of students, mainly women, turned up in the institute wanting instruction, the scientific drive was feeble.

Now rested, Jakob began to busy himself with a galvanometer and other apparatus on his laboratory table. It occurred to him that he might be the only physicist in

Germany, perhaps in Europe, who at this moment was preparing to make a measurement of a quantity for no other purpose than to know it precisely. In this way he might help keep alive the spirit of physics, but, seriously, could anyone regard this puttering in the laboratory as the fulfillment of a duty? Wasn't he elevating to a principle what was merely a pastime? This question was hard to deal with while he was trying to decide where this and that wire went. The answer seemed to come not from him but, objectively, from the laboratory: it was egotistical of him to believe that his handful of wires constituted European physics at this moment, that his small measurement had any significance for the scientific spirit. Besides, physics didn't belong in this laboratory any more, but out there, where nations were fighting for their lives. Jakob laid down the wires and walked to the windows, which he more looked at than through, since the lights were on inside and outside it was overcast. As in a mirror, he saw the untrimmed mustache, the bony nose, the furrowed skin receding over the top of the skull, the wisps of gray hair around the collar.

There were times when he found physical deterioration in old age objectively interesting, but this was not one of those times. In any case, what was important about a man like himself was not physical but mental. If he were asked what kind of thinker he was, he would say that Ostwald's distinction between classical and romantic temperaments in science applied to him. Though he was no genius, he saw a similarity between himself and physical thinkers he admired such as Helmholtz, Gauss, and Gibbs, who were classical in Ostwald's sense of the word. Like them, he insisted on building physical theory from the ground up, and he held back publication of his work to make it better and better. He didn't know if he had fewer and slower ideas than his temperamental opposites were supposed to have, but he pursued what he had with classical tenacity. Naturally he didn't match Ostwald's type in every respect, but there was a fair correspondence.

Jakob was struck by the differences when he compared himself with Des Coudres, the embodiment of romantic temperament in physics. To begin with, Des Coudres lacked the patience to put his thoughts in systematic order or, for that matter, even on paper. Jakob wrote everything down and numbered the paragraphs. Des Coudres had even fewer publications than Jakob, though he had done better in the world as professor of theoretical physics at Leipzig with an institute of his own. His bent was to subject nature to unprecedented distortions, which was not at all Jakob's. Des Coudres and his students built bombs to withstand enormous pressures, and with them they studied matter in extreme states in which liquid and gas were one, as in those once fashionable paintings of sea and mist without horizon. Des Coudres was excited by the world of atoms and everything that was new in physics, things that so often raised Jakob's hackles. It was the same in art, where Des Coudres followed everything new, no matter how disturbing it might be. Jakob wondered how Des Coudres could listen to Wagner's *Walküre* three hundred times! Vacations sent Des Coudres to the great works of art in Italy and then farther south. On colorful stationery provided by steamer lines and hotels set amid palms, Des Coudres wrote to Jakob from Naples, from Jerusalem with its colorful mix of peoples and settings, from exotic Port Said. It was in character that Des Coudres should volunteer for the front as medical aide at the start of the world war and that, when his tour of duty was over, he should want to stay. He wrote to Jakob that on moral grounds he wanted to stay at the front until the war ended, since he could do more that way for the fatherland than by facing meager audiences at his university.

Jakob didn't return to the laboratory table to complete the measurement, which could wait. Instead, he set off to visit the Geheimrath, whom he used to find down among the fragile and costly pumps and glassware in the low-temperature laboratory, but more often than not he now found in his office. Two years ago the director had begun to do war work in the institute and soon was doing little else. At the time, he was furious with protesters, who made people nervous about the leadership of the war and who were totally ignorant of the military matters they attempted to judge. When the director drew up his own protest to make this point, Jakob remarked that it wouldn't do any good to keep ignorant people from judging military matters if other ignorant people like themselves proceeded to do just that. So when the director's protest was mailed to professors and businessmen for their signatures, Jakob's name wasn't on it, and from that time on he felt that the director regarded his patriotism as deficient in ardor.

As Jakob entered his office this morning, the Geheimrath was hardly visible at his desk behind piles of loose papers, reprints, instrument catalogues, and pipe fittings. Irritably, the Geheimrath looked up. Both of them might as well go back to bed, he said, since almost everyone was away or dead, and now there was even talk of closing down the university. He began to complain about his life, about being too young in 1870 and too old in 1914, about the waste of a good cavalry officer. As a physicist he had tried to do some good for his country. Whenever former students wrote to him for help with military problems at the front, he dropped everything to work on them. It was the same when the army or navy called on him. But no matter how hard he worked, he was only one institute director. What was wanting was organization. There was a beginning with Max Wien's physicists who designed wireless telegraphs for pilots, with Rudolf Ladenburg's who designed ranging instruments for artillery, with Gustav Hertz and other physicists who worked with chemists on poison gas. But altogether too much physical talent was idle, and physicists like himself who tried to coordinate efforts were told to relax, the war would soon be over. Well, they were wrong, and if physicists had been organized Germany would be in a stronger position today. After all, German physicists were imbued with the special German drive toward the foundations of knowledge. They were not content with brilliant successes based on chance but sought the slow, difficult way, mastering every weakness as they went along. Properly organized they would have made a difference. The Geheimrath had given a great deal of thought to the German genius and the war.

Jakob asked about all the loose sheets of paper he saw covered with penciled arrows, ladders, boxes, balloons, and the like. The Geheimrath explained that they were schemes for permanently connecting physical institutes with the military forces. He wanted to see institutes built right into mobilization plans for wars to come, which would prevent the delays and disorganization of this war. Professors and officers would keep in touch on military needs and research in the future. Rising now to stand beside a photograph of himself on a splendid mount, the Geheimrath explained that he had been trained in classical warfare and knew how to maneuver a squadron in front of an advancing army. Opening a book on his desk, he showed

Jakob photographs of Flanders. They were the usual ones of trenches full of tired soldiers in field-gray, of bored and dazed troops massed at rail depots, at bivouacs, at hospitals. The Geheimrath turned pages until he came to a picture of a smiling young aviator with one hand resting lightly on a wing over a rack of bombs. There was one fighting man with skill and verve, the Geheimrath said brightly. What caught Jakob's eye was the picture on the facing page of a plane falling straight down, trailing smoke. Above it were two tiny billowing shapes, thank God. Did the Geheimrath know that pilots didn't last a season, Jakob asked. If people must be killed, let them be people who court danger, who come twelve to the dozen.

As usual, the institute library was empty, or almost empty. Hidden behind a stack of thick manuals and bound journals was the first assistant of the institute, now the only assistant. In his years at the institute, Jakob had seen many assistants come and go. There had been the subservient assistant, the arrogant assistant, the assistant who would never accomplish anything in physics, the assistant who predicted that his own name would live forever and Jakob's would be totally forgotten (for some years Jakob watched anxiously for the man's name in the literature), and the many conscientious, overworked assistants who made the institute run well. Jakob missed the Russian assistant whom the Geheimrath immediately dismissed because he was an enemy alien and the Swiss assistant whom he dismissed soon after because he wasn't German, which Jakob didn't think was a good reason.

The first assistant came up to him now. If the professor had a moment, he would like to talk with him about a subject he knew was close to his heart, the world-ether. He had been reading Mie's stimulating papers on the theory of matter, which went beyond Einstein's general theory of relativity and which could be brought into relationship with it. It was exciting to see how it could be done, and for that purpose one had to read Hilbert's recent work on the subject. The professor followed him to his seat and with sinking heart looked down at the densely massed equations filling the open pages. It would take a lot of mathematical power, probably more than he had. The assistant stopped running his finger along the equations and gave the professor a friendly look. Reflecting how really indispensable this assistant was to the institute in these times, he was sorry that he still had doubts about how the assistant would work out in the long run. The trouble was that the assistant might alarm certain people in the institute who believed they had experienced Catholic chauvinism at one time or another. Unwisely, the assistant had joined a Catholic society, which could draw official displeasure onto the institute. Relationships within the institute and with the ministry were delicate, and the professor would have to have a word with the assistant one day about that society, really in his own best interest.

But he had come to look at recent journals, the few that the institute still received in the war. From long habit, he looked to see if there was anything he should be aware of for his lecture this afternoon. There wasn't, of course. No one was interested in pure acoustics anymore.

As he stepped out of the library, he turned right, only to spot the custodian of the institute lolling in the corridor. He wheeled to the left, even though this meant walking out of his way to his office. He would go to any length to avoid the custodian, whom he spoke to these days only if he had a witness along. He must have another word with the Geheimrath about this man, though he doubted that any good would come of it.

Back in his office with an hour to go before his lecture, Jakob thought about the mathematical methods of physics. Over his lifetime, physics had taken a turn toward increasingly advanced mathematical conceptions of nature. Fifty years had not proved long enough for him to see into the depths of the equations of classical physics, certainly not into the final revelations of Maxwell's equations, which Hertz correctly saw were wiser than their creator and his followers. And it was unrealistic of him to expect to see into the depths of the equations of physics that came after Maxwell, if they had depths and were not a mathematical trick in the end. (For weeks he had been struggling with a paper on atoms by Sommerfeld, only to conclude that he was not doing physics but conjuring with numbers.) He was a sad contrast to the assistant, who evidently was not deterred by the mathematics of relativity. In the past, mathematics and physics had had a close relationship, but Einstein fused the two, and Jakob found it hard to see where the physics was.

In fact, theoretical physicists' whole way of speaking about the world was shifting toward the mathematical, a development Jakob didn't welcome. For mathematicians it was proper, but for physicists it was dangerous to speak of *curvature relations* of space, for example: that was to avoid speaking of physical conceptions, as if the physical problem were solved within the mathematics itself. Jakob knew that half the problem would be solved if certain physicists were not determined to forget the word *ether.* Without greater mathematical knowledge, Jakob stood little chance of releasing certain of his dark presentiments about the world-ether. It was a loss.

He thought of the mathematicians' opposite numbers, whose direction he liked even less. They confused touch, taste, and feel with thought, and they all but dismissed theoretical physics. Whereas he could agree with Boltzmann that physicists saw the methods of theoretical physics subjectively through their own spectacles, and whereas he could agree with Volkmann that a subjective element entered with the understanding of theories as only pictures of nature, he drew the line at Gehrcke's claim that personal sympathy and antipathy were the foundation of all theories in physics. It could only mean that physics was not rational and that physicists lacked the ability to see viewpoints other than their own and to judge how far their thought agreed with reality. It was in keeping with Gehrcke's misunderstanding of the nature of theoretical physics that he should accuse Einstein of lacking all physical feeling, of standing outside phenomena instead of grasping them from within, of crippling the imagination of physicists. Einstein's work had to be criticized—Jakob had his criticisms—but, as Born said, Gehrcke went outside the limits of scientific criticism when he denounced relativity theory as a mass suggestion, likening it to those nonexistent N-rays the French tantalized the scientific world with a few years before. Jakob was especially irked to see Gehrcke repeat this nonsense in his posthumous edition of Drude's classic work on optics. Drude would have been no partner to it, Jakob was sure.

He checked his watch. There were still a few minutes before his lecture was to begin. He closed his eyes and summoned up pictures of Greece. Home of rational thought. He was standing alone, holding an orange. From where he was, high and clear, what he could see was strangely abstract and alienating and at the same time luminous, alluring. It was an infinite horizon, a pure Euclidean division of infinity, uncluttered by objects and people, a pure nonverbal radiance of the reasoning mind. He might have been looking at the inner world of thought and the outer world of the universe, each an infinity. He might even have had an intimation of total understanding. For mathematics ordained laws that guaranteed the conformity of human thought to the natural order. Mathematics, the Olympic patron of theoretical physics. He bit into the orange and sucked on it.

Jakob sat up suddenly, feeling mildly discouraged as he did after returning from reveries like this. Their recurring image of an infinite horizon shamed his actual mathematical understanding. He checked his watch again and saw it was time to go. As he left the office, he already felt better and was looking forward to today's lecture on acoustics, one of his favorite subjects. With its clear and precise mechanical principles, it was classical physics at its best. Moreover, in the most direct way the phenomena of sound revealed to the senses the harmonies of the physical world, which was evident from the names of instruments on the acoustical table, the harmonograph, the sonometer, and so on.

He entered the lecture hall, his spirits rising. Then he saw the blackboard. The blackboard!

It was not clean but written all over with the Privatdocent's formulas. That was irritating enough, but the formulas were irritating in their own right since they belonged to an atomic physics that he was skeptical of and that didn't belong here in any case. He would have turned on his heel to call the custodian, but the custodian was in collusion with the Privatdocent. Angrily, he swiped at the blackboard with his handkerchief, leaving a white smear across the offending formulas.

Shaking chalk dust from his handkerchief, he looked up at the scattering of students. Balls of sweaters, coats, and scarves on this cold afternoon. In his mind's eye, he saw the hall before the war, when there had been so many students that they sometimes had to stand. With the windows closed to keep heat in and noise out, the hall had had an intimate quality then and he had felt good about his lecturing. (He had got an income from student fees then, too, that wasn't so negligible as now.) As he began to lecture and students bent over their notebooks, he saw massed crowns of hair, women's. He didn't know these students.

Jakob began: They would remember that mathematical physics began with music and the Greeks, with Pythagoras. Mathematical physics long ago separated off from music, but the original inspiration of music to scientific thought persisted. Without it, he told them, Kepler wouldn't have discovered the harmonies of the planetary world on which Newton's work and with it all of modern physics rested. So if they reflected on it for a moment, they would see that without music and the exact way of thinking about the world it inspired, this physical institute wouldn't be here and neither would he or they. (He might have qualified this generalization about the musicality of physicists, since he was all but tone-deaf. Like Hertz.)

The professor explained that sound was caused by a wave sent out from a sounding body, as he now demonstrated with the siren on the table. By turning the crank steadily he produced a note, and by turning it faster—he was breathing quickly now—he produced a higher note, which showed that the pitch of the note corresponded to the number of waves reaching the ear per second. The students had laid down their pens and looked sleepy.

Jakob tried to be a good lecturer, tried at the very least to keep his students awake and learning. He presented them with phenomena, laws, proofs, pictures, history, problems, demonstrations now and then—and he used voice, hands, chalk, projector, exhibits, instruments, numbers, symbols, equations, charts, graphs, geometrical models, and anything else that worked.

Jakob had given much thought to the art of lecturing and never more than in that year when he temporarily directed the institute and gave the big experimental physics lectures. They were the only physics lectures for most students, who would soon go off to cure sore throats or souls or sit at desks over ledgers, and Jakob wanted them to count. To learn how a master did it, he went to hear Lehmann at Karlsruhe. Lehmann's lectures were scarcely lectures at all, but revelations. Not restricting himself to the demonstration table, he used the whole floor for his experiments, working them by powerful engines beneath trapdoors. He kept two assistants' hands full and a vast audience of beery students coming back. His performance was not within Jakob's reach; so he went to hear the legendary Warburg. So smoothly did Warburg lecture that he might have been reading from notes, and while he lectured he connected wires and gas jets, slid weights, joined pipes, read meters, and worked all manner of complicated apparatus. This, too, was beyond Jakob, and so it was by instinct and hard work that he got through the year of experimental lectures. It was with a certain relief that he went back to his small theoretical physics lectures after a new director was brought in.

So his success didn't depend on his skills at the demonstration table, by and large. Nor did it depend on a perfect memory or a projecting voice, neither of which he had. Nor on a world reputation, which made up for a lot of deficiencies in students' eyes. It didn't depend on technique at all, but on the educational value of his subject. Theoretical physics offered students precise and fundamental knowledge of the external world and of its relationship to the internal world. That was a high order. Jakob felt elevated by his discipline.

He clapped his hands sharply. Cannon, he began, have advanced the course of science more than the course of nations. They have provided admirable signals for measuring the velocity of sound.

He explained: Like every other part of a measurement, the measuring observer enters an equation. In measuring the velocity of sound, each observer has a different reaction time for recording muzzle flashes and powder claps, so he must take his personal equation into account. (Yes, he resolved, in my coming talk, *with the precision of a personal equation* I will describe the thoughts and feelings of a German physics professor at a time of upheaval in physics and in the world.)

He digressed: Cannon commonly damaged ears in war. Passive, helpless organs, ears can't be closed, and it doesn't help to cover them with your hands. Better to open your mouth to equalize pressure. Better yet to use an antiphon, which remains to be invented. He looked up at the periodically jagged flames projected hugely on the screen. He explained that they were seeing pictures of flames produced by a burner in which the flow of gas was modulated by a membrane vibrating with sound waves. They were *seeing* sound.

He digressed again: When the history of this war is written, he said, classical physics will share the credit for its conduct. If, God forbid, the world should go to war again, it will use modern physics. But *this* war was fought with *classical* physics, with every branch of it. To locate an enemy gun, for example, they used methods from optics, electromagnetism, seismology, and other branches of physical science, including acoustics. To locate a gun acoustically, they compared the times of arrival of the sound of the gun at several observation posts, for which purpose they used fine Swiss stopwatches that measured to one-fiftieth of a second. The professor looked at his own watch, always fast. There was still time.

The war front was a vast stretch of unpredictable weather, wind, din, flame, and smoke, and enemy gunners had firing tables of their own and were using them to try to kill our gunners. It was remarkable that physics could be applied in battle at all, for it normally described the simplest imaginable events, observed under the most favorable, controlled conditions. In war, events were the most complex imaginable and conditions of observation the worst. The professor didn't go on, since the subject of methods of approximation quickly lost contact with action and his audience's concerns. He snapped off the projector light, extinguishing the flaming horizon.

Next he wrote formulas on the blackboard for a while. As he did he reminded himself that the war front was a network of ears, eyes, brains, acoustical rangefinders, telephones, telegraphs with and without wires, periscopes and stereotelescopes, heliographs, and many more devices. Looked at that way, the war was an abstraction, much as it was in newspapers reporting that the Western front had shifted so many kilometers to the west—or, lately, to the

east—and that so many enemy panzers had been destroyed and so many tons of enemy shipping had been sunk. Jakob keenly regretted his lack of living connection with the war, the greatest event he was likely to live through. Early in the war, his younger colleagues and students often sent him field postcards. Like tourists' postcards before the war, they showed pictures of churches, only these were bombed out except for an arch or two, a piece of nave, a few upright tombstones. Sparingly, the postcards reported hard marches in foul weather along bottomless paths, bad lodgings, soggy trenches, rats, the dead. Jakob wrote back long letters about everything that went on in the institute, momentarily redirecting their thoughts toward the high tasks of pure science. He also advised them to enjoy the life of war if they could and then to return with good conscience at having done their duty (and to bring back any pictures they had of the battlefields). It was at this time that he had walked with a group of Sunday tourists through a little practice trench to feel what it was like. That had been early on, and he was embarrassed when he remembered it now.

He didn't lecture on the physics of the war, a theme that could easily fill the lecture hall. Auerbach had published his Jena lectures *Physics in the War*, which had already gone through four editions and which Jakob warmly recommended to his students. In it, they would find full descriptions of the current technology of war, with plenty of pictures to prove that physics was an inexhaustible source of that technology. Although war was the worst of trades, it was a trade all the same, and so it must draw on the science of the forces and energies of nature—on physics. Today, physics served the cause of destruction, but soon the gardens of civilization would replace the deadly battlefields, and then physics would be cultivated for its own sake as it had been in the past. That was what the professor told his students, and he believed it, almost certainly he did.

It was irritating to have to write over the messy Privatdocent's formulas. The professor's chalk broke frequently.

At the end, as students filed out, Jakob stepped back to admire what he had written last: the concise equations of an ideal mass point moving under the influence of elastic forces. How beautiful classical physics is! (Was! He meant before the quantum theory of the atom.)

Before he left the hall he wrote in block letters: DO NOT ERASE! He had had words with the Geheimrath before about the disruptive influences within his institute. For his trouble, he had been reminded that the Privatdocent needed a lecture room and that he had generously carried out Jakob's duties during his illness. Besides, the Privatdocent had returned from the front with nervous disorders, and Jakob should find it in himself to show tact and feeling. So it had come down to a struggle over the blackboard and the Privatdocent's always repeated (and physically dubious) formulas. Jakob thought that this behavior had stopped, but today he saw it hadn't.

The Geheimrath was where Jakob had found him this morning, at his desk, calculating for the war. There was a point at which conscientiousness bordered on morbidity, Jakob thought. It seemed unfair to him to burden the Geheimrath with personal problems. So instead of complaining about the blackboard, he encouraged the Geheimrath to talk about his lectures, which his publisher wanted him to prepare for publication. The Geheimrath had started to work on them last winter, but dropped it when the navy asked him to work on U-boats. He thought it would be only for a short time, but now the war had grown unpredictable again. Besides, he found it hard to put himself into a physicist's frame of mind, since an inner voice said to him that this is war, that everything else is meaningless—including physics. Jakob conceded that not much physical research had been done in recent years when most physicists were away at war and that physics journals were starved. But some research got done, and it would leave its trace on the history of physics. That was the whole point. No war, not even this catastrophic war, no numbers of submarines, planes, and tanks could long discourage the common quest for understanding the physical world. It was not meaningless!

The Geheimrath seemed not to have been listening, for he suddenly asked: Who can make sense of Bohr's atomic theory? At one moment an electron was supposed to be on one path and at the next moment on another path, and the theory couldn't say how it got from one to the other. The theory said only that the atom sent out a pulse of light with given energy and wavelength when an electron sprang from one path to another and that it didn't radiate at all when the electron was in its stationary path. The theory was nothing more than rules for calculation, which provided no understanding of atoms. Jakob had sympathy with the Geheimrath, and he also had sympathy with the developers of atomic theory because of their seemingly impossible task. He said that it was to Bohr's credit that he could imagine a structure for something that was unimaginably small and yet was capable of countless periodic motions. The difficulty was that the inner workings of Bohr's model of the atom were governed by hypotheses that no one would take seriously if the model weren't capable of accounting for certain facts surprisingly well. Now Jakob accepted the need for hypotheses, for otherwise physicists couldn't treat invisible things like atoms, which certainly existed. What he could not accept was that atomic physicists were permitted to begin with any hypotheses whatsoever and to proceed to calculate a world with virtually the same freedom as mathematicians. That was happening these days, it seemed to him. It set his teeth on edge. Convinced that nothing in physics required riper experience than the introduction of proper hypotheses, it hadn't escaped his notice that most atomic theorists were near beginners in physics.

At the end of Jakob's passionate speech, the Geheimrath looked unhappy and wondered aloud if theorists were necessary to physics at all. He had often noticed that they thought of themselves as mathematically abstract thinkers who were mentally superior to experimentalists. Theorists even managed to persuade some experimentalists of their inferiority, so that they only presented their work in the shadow of some great theoretical name. Experimentalists working in laboratories created all that was new in physics and theorists merely rewrote their work while sitting at desks. The Geheimrath was staring hard at him, but Jakob had heard these severe opinions of his many times before. Jakob explained that the experimentalist's main task was to test theories and measure their basic constants, whereas the theorist's was to reduce phenomena to the simplest laws. The Geheimrath brought up the complexity of phenomena, which experimentalists understood better than theorists because of their practical experience. Jakob pointed out that the history of physics showed that connections in nature were less complex than previously imagined. So they produced argument and counter-argument, as they often had in the past, until Jakob was inspired to compare the conflict between theory and experiment to the conflict between school and life, between learning and experience, which from a higher standpoint was seen to be no conflict at all. The Geheimrath then conceded that Bohr's atomic theory for all its shortcomings had given experimentalists work to do and that the exact measurement of Planck's quantum of action was an important task of experiment. Jakob then conceded that it was precisely the new facts that experimentalists turned up that forced theorists to modify their views. The Geheimrath admitted that when he began in physics he could understand its explanations, but not always now. Jakob admitted that much of modern physics passed him by too. He doubted that the quantum theory, strictly speaking, was a theory at all, and he even allowed the Geheimrath to doubt that Einstein's general theory of relativity was a physical theory. The Geheimrath had an arguable point: experimentalists had to be able to understand a theory before they could test it, and an untested theory didn't belong to physics.

The Geheimrath began talking about U-boat wireless telegraphy and about enlisting the British physicist Maxwell to defeat the British navy. Jakob left quietly.

Having never experienced the custodian's curling lip, the Geheimrath had no idea that Jakob was being persecuted. Only the government could fire the custodian, which wasn't easy if the case presented itself, so to speak, as sociopolitical in nature. It did no good for Jakob to argue that it was a simple case of justice demanding discipline. Having grown steadily more insolent, the sloe-eyed custodian had finally unmasked himself by spreading rumors about Jakob's behavior toward young women in the institute. Jakob lost sleep brooding about the custodian, who was usually seen leaning on his broom, gawking and making lewd jokes. Jakob's illness at the Patriotic Evening was surely brought on by nervous strain aggravated by this institute affair. Back in his office, he wrote a firm note to the Geheimrath: the custodian must be fired or he will gain certain tyranny over us. He has regularly lied to get his way, betrayed at least one of us by vicious gossip, defied orders in every conceivable manner, insulted feminine honor, and has even tried to lay hands on a young woman in the institute. I can't urge on you too strongly the need to inform the ministry, for every day the man remains, work in the institute is obstructed. The ministry should be reminded that since professors are preoccupied with scientific teaching and research, they ordinarily brush aside minor unpleasantries as they brush aside gnats to ward off bites. But they are now plagued by a swarm of gnats, requiring mosquito netting . . .

Before leaving his office, he looked in the adjoining room where he had begun a measurement earlier that day. It was good he did, for the laboratory table was not as he had left it. The galvanometer, his own, was gone, and no one but the custodian could have taken it. Jakob had lost parts of apparatus before and had confronted the man, who denied everything and was hostile to Jakob ever since. It was hard to believe that the custodian stole from the institute. Jakob would demand an accounting once again. He sighed.

Jakob was almost out the door of the institute and on his way home when he had a premonition. He turned back to look in his lecture hall and confirm his fears. His own incompleted lecture, not to be erased, was erased and in its place the Privatdocent had put back all of his formulas. More upsetting than that, in a lower corner of the board the Privatdocent had drawn a small square. Inside it, in his undisciplined scrawl, he had written in *red* chalk:

PROF. JAKOB'S SPACE

No, it was his space, all of it!

Not knowing exactly why, he returned to his office and took out the revolver he had kept—and forgotten—in the bottom desk drawer.

Crack! Crack! He remembered how at the beginning of the war he had stood in the garden of the physical institute with a smoking revolver in his hand. What in the world could he be up to, they wondered. Crack! He saw the custodian and the porter draw their heads back behind the window casing in the lecture hall. The simple explanation was that he was making an experiment on sound that was going to help win the war. At his post, everyone did what he could.

Jakob stuck the revolver in his pocket and started off, now for the second time.

Crack! Branches the wind had stripped from trees broke under his heel. Walking along the path outside the institute, he thought he saw his shadow cast onto the fog ahead. He shivered from the damp and cold. Like a family, people in an institute get on one another's nerves, he reasoned.

Soldiers were idling in the institute garden, not deserters he hoped. What were they doing so far from the tavern? Jakob heard a woman laugh—at him, no doubt, an old professor shuffling by—and he couldn't help comparing these soldiers with soldiers in the past. But why should he expect discipline in soldiers today when there was so little to be found anywhere? They had probably picked up their ways in the Youth Movement. They abused culture, and they knew nothing better than to sing songs about nature and freeze all night on hikes because they didn't know any more about nature than about culture.

Anxiously, Jakob hurried from the institute, past open fields, past woods, past the new glass factory, which was still without glass in its windows. Its construction slowed by war shortages, the gaping factory stood there ugly and disgusting. Now Jakob marveled at industrial technology, which showed humanity's ingenuity, its boundless capacity for organizing nature for its convenience, its synthetic imagination. Indeed, modern Germany was inconceivable without it, and he knew that it together with physics and the other natural sciences was responsible for changing the material conditions of life, freeing people from drudgery, awakening their intellect, and promoting personal and civil freedom and moral excellence. So he was annoyed to hear those pampered vagrants from the Youth Movement reject the industrial age and call for a return to tradition, community, Germanic primitivisms, and other varieties of unreason. He was annoyed even more to hear his colleagues at the university reject it or at best grudgingly acknowledge its slight alleviating effect on the suffering masses. If once, a thousand times he had heard the litany of fashionable complaints: telephones, gramophones, airplanes, and like inventions created distractions and caused the population to increase through the multiplication of material things, with the result that people needed electric lights because they were crowded into cities, which did nothing for their well-being . . . He saw danger in this facile rejection of the modern world. He admired Nernst, who volunteered in the war and drove his own car close enough to Paris to see its lights before the Germans retreated. Yet in his heart he was not as close to the technological, to the man-made, as to the natural. On mental finger tips, he enumerated his preferences: towns over cities, fields over factories, leisurely walks over motor holidays. So at the same time he welcomed the glass factory, he foresaw the pall of smoke it would lay over the university. After all, universities were one of the last sanctuaries in the modern world where one could reflect on humanistic ideals. In truth, he felt profoundly ambivalent about the factory.

Jakob stopped and sat down on the ground, resting his head against a pipe intended for the factory. Despite the cold, he might have fallen asleep if the Geheimrath hadn't pulled up honking. Jakob climbed into the gleaming machine, glad for this mechanical assistance. The Geheimrath was returning from the small airfield outside town, where he had been given a ride up by pilots from his class on aerodynamics. He told Jakob that when he retired he was going to own his own plane. What did Jakob think of that?

Jakob told the Geheimrath about the time he had gone up in a plane. They fitted him out with helmet and goggles so that he looked just like an air ace. They took him up high over the town, 2000 meters, 2300 meters . . . when suddenly the plane dropped down to housetop level. He let go of the battery lamp with its convenient button for signaling, which sailed off into space and spoiled the war experiment.

The Geheimrath asked him what his sensation had been. Jakob answered that he experienced the normal direction as the resultant of the gravitational and centrifugal forces. The world seemed to be standing at a tilt!

That was not fantastic, the Geheimrath shouted over the noise of the motor. Did he know that when planes were as safe as cars, there would be regular flights everywhere, and airfields would be connected to inner cities by trains?

They passed a field with cows and then a young woman carrying milk. Jakob could read her armband, Children's Milk Establishment, which reassured him that the car couldn't be moving too fast. The Geheimrath was paying no attention to the road at all as he commented widely on affairs.

The Geheimrath recalled that before the war he had urged the Germans to build up their air power to keep up with the French. He foresaw that war planes would fly behind the lines to destroy railways, electric stations, war plants, wharfs, supply columns, troop positions, and headquarters. Events had proved him right, but too late. The development of air flight for war uses had been slowed by the dream of permanent peace, which every prewar scientific congress had encouraged. War was barbaric, but they were still living in a barbaric age. Each people must earn its right to exist by force, as Babylonia and Egypt forgot. Fortunately, Bismarck remembered. The essential thing was that the higher cultural races should act responsibly by not interbreeding with lesser races, which would only produce a lower type. Germany needed good organization and the willing sacrifice of the individual for the whole. Wisdom couldn't be separated from might.

Without warning, the car ran into a ditch, tossing the occupants out gently, like apples rolling from a basket. Jakob didn't move from the muddy ditch where he landed until he was certain of his limbs. The Geheimrath had disappeared under the overturned machine to find the trouble. The bolts in the steering column broke, he explained, and we lost our steering!

But Jakob was hurrying down the road, and soon he couldn't hear the Geheimrath banging on metal.

Soaked through and caked with mud, Jakob normally would have felt sorry for himself. But his apartment building had just come into view, and his thoughts turned outward. The building was a mixture of popular historical styles of the day, a touch of Florence here, Venice there, Rome there. To Jakob, it was painful, like looking into a sandstorm. Like most recent buildings, his lacked a coherent idea, and in that respect it mirrored the age. He thanked his sound education in classical antiquity for the standards he held on to or, to be precise, clenched between his teeth. Modern culture was kitsch. It held to nothing absolute. It lacked—he sought the word—discipline, which brought his complaints around to that austere preserve of the disciplined virtues of hard work and rational thought, his own theoretical physics. Even that bent to fashion and had been bending for thirty years. If any of the young discoverers dreamed up a hypothesis about the atom or the universe, he rushed to print with it. He was encouraged to if his hypothesis violated every canon of theory construction. For his audacity the discoverer gained notoriety and imitators for a season, and then he was forgotten in the next enthusiasm. Instant publication of research was the rule, and it didn't matter if it was only half thought out. Teaching also suffered, for good teaching took as long to ripen as good research. The truly original personality in physics could emerge only under proper conditions, which for the old professor meant past conditions. Physicists used to reflect on scientific problems until they penetrated to the depths and exposed the inner connections. That was in a more serene time.

He was an appalling sight as he stood there in the entrance of the apartment, a puddle forming at his feet. His wife helped him out of his dripping clothes and into his robe, saw him into his study,. and disappeared to make something hot to drink. He should never have left home in the first place. Now look!

With a blanket wrapped around his legs and a cup of tea warming him, he felt better, even a bit drowsy. His thoughts returned to factories, airplanes, and autos, and he tried to put the ambivalence he felt about these things into perspective. He was resigned to the perpetual confusion between scientific discovery and electric trams, zeppelins, and the whole panoply of technological feats. He didn't deny that physics owed to technology its instruments and certain scientific problems. Nor did he deny that physics gave rise to technology, as he had told his class just this afternoon. Helmholtz had got it right when he said that the entire recent development of industry depended on the mastery of natural forces, which the laws of physics gave us. Yet Helmholtz knew that physics was not pursued for utilitarian reasons but for its own sake, freely, idealistically, as it had to be if it—and with it technology—were to advance. The physicist sought an understanding of nature, and the applied physicist and engineer and technologist turned that understanding to practical benefit.

Jakob used to be afraid that if technology got a foothold in universities, it would subvert their idealistic pursuits, that its trumpets would drown out the subtler harmonies of the physicist, and so forth. His fears proved groundless, but he still believed that technology was best left to its own institutions.

Jakob was annoyed when technologists boasted that they had given physics and other natural sciences so many stimuli lately that the center of gravity had shifted from science to technology, that technology and not science determined the course of the modern age, that scientists like everyone else now thought technologically. But technologists did not alarm Jakob as they once had. In seeing themselves in a superior light, they were only typical of all specialists.

In the same way, artists sometimes got on Jakob's nerves. Instead of living at peace with scientists and recognizing their view of the world, artists challenged scientists and insisted on their superiority over them. Jakob thought of the circle of artists in Munich who had recently resurrected Goethe's theory of colors and confronted the physicists with it. Their foolish talk about the physics of color was of a piece with their nationalism, which fed on their hatred of the English, including Newton with his theory of colors. Sommerfeld had finally exposed them, though Jakob suspected it wouldn't make any difference. According to these artists, university physicists had no right to decide between Goethe and Newton because they were partisan, as if scientific truth were a political persuasion. If Jakob appreciated Goethe's researches on colors, it was not for their physical ideas, but for their loving observation of nature and their psychological fidelity. (He noted that both of Helmholtz' old essays on Goethe had been reprinted, and he had seen a new book by Ostwald and other recent publications on Goethe's theory.)

Trends in the arts in the twentieth century left Jakob cold. He had respect for art, of course, and its just claim to share with science the pinnacle of culture. What he couldn't abide was the celebration in the name of art of all that was mystical and irrational. There were people who accepted science only insofar as it was subjective and dealt in value judgments, which denied culture its objective and factual content. They wanted an education that was wholly artistic, which, Jakob knew, would produce a mass of frivolous dabblers since artistic talent was as rare as scientific talent.

Jakob had no patience with Warburg, who talked of physical theories as one of the highest forms of art, as artistic pictures the soul created of the inconceivable, as descendants of myths of the creation of the world. Evidently, some physicists didn't understand the forces that worked on their thinking, and so they had to be told. So Simon told Warburg, and so Jakob would like to tell Wiener, who claimed that the poet and the scientist were sparked by the same fire, that poetry was present in science no less than in the works of poets.

Poetry and science were different, as poets knew if physicists didn't. Otherwise, why would poets portray science professors as monsters of specialized pedantry? He knew science professors, and he knew that they were not odious beings sunk in uncreative materialist delusions. They had their limitations as well as their strengths, which was what could be said of poets or any other specialists.

Of course, like technology and art, science had its prophets, which did science no good. The Social Democrats preached that Darwin scientifically proved the workers right, and the Monists turned science into a religion. As Jakob saw it, Monism was a parody of modern biblical scholarship, which all but reduced Christianity to a set of ethical prescriptions useful to society. From this, as from most trends, Jakob dissented. He regarded himself as a good Christian, which meant that he believed in God and in his calling, physics. He didn't go to church, since what he heard from the pulpit usually annoyed him, and naturally he didn't go to those sham scientific devotions, the Sunday Monist sermons.

Jakob raised himself on an elbow to write in his notebook. He felt the need to control the rush of thoughts by fixing them on paper. Specialization, he wrote. It was obvious that people who devoted themselves to technology, art, or science—or to religion, business, or the military—often misunderstood one another, simplified the world, and claimed it for their specialties. Jakob was himself a specialist, a physicist, a *theoretical* physicist, and at times he believed that theoretical physics had generated all the natural sciences and even rational thought itself. So he knew the dangers of the specialist's way of thinking. But at the same time he was appalled by the fashionable ranting against the degradation of learning to bread-and-butter study, against details and facts, against the unhealthy splintering of knowledge, against specialization as the sickness of the age.

Really, these complaints belonged to cranks (Jakob had

received letters from a few genuine cranks) who claimed that our ideas of nature were wrong because natural science was divided into physics, chemistry, and the like. They overlooked the need for specialized disciplines in the face of the quantity and complexity of knowledge. In fact, the need to specialize had advanced so far that physicists in the old sense of the word hardly existed. Instead, there were optical specialists, electrical specialists . . . Even physical institutes were becoming specialized institutes for spectroscopy, low-temperature work, and such. Certainly the day was past when an individual could unite several sciences and their applications, as Gauss had done in the mathematical sciences. Now specialists came together in shared work and institutions, as at Göttingen where mathematicians, physicists, and astronomers recognized their science as a totality like Greek culture. (Jakob suspected that Göttingen had nothing to do with Greek totalities but was only a further illustration of his idea that every specialist, here the mathematician, expanded his science to encompass everything.)

Suddenly, the study door was flung open. Jakob couldn't believe his eyes. There in the doorway his wife stood, looking furious and holding a revolver in her hand. Where had the professor got this thing that had fallen from his soggy coat? What had he intended, what possessed him? It took him some time to explain and to get the revolver back from her. Embarrassed, he quickly removed it from sight under the pillow on the couch.

The pillow was hard under Jakob's head, and he reached under it to shift the revolver. Tomorrow he would return it to the institute. To the institute . . .

The professor heard out the Geheimrath, who was wearing his reserve cavalry boots and student colors: I am the director of the institute! The Geheimrath sounded and looked splendid. I alone have every detail in view. Everyone reports to me and asks my permission. My subordinates—you Jakob—would not exist as physicists outside my institute. Without me you would be . . . Tell me the word, Jakob. Chaos! You, all of you, would wander about like gas atoms. I am here to prevent that, to tell you what you can't do. I am the negator of waywardness, the lawgiver, the guarantor of the whole. I am accountable to no one in the institute, to no one in the university, to no one —except the ministry. I tell you all of this, Jakob, because I detect in you a certain independence, a tendency to think you have a right. To avoid collisions, at great labor I've written out for you the laws of my household. The assis-

tant, porter, and custodian are mine. The space is mine. The instruments are mine. That is how we avoid collisions. The professor protested: If I have to get your permission every time I need a thermometer, my ideas will all disappear. It's unheard of for a director to make no distinction between an ordinary (even if honorary) professor and a beginning student. The Geheimrath explained: The ministry is behind me in all of this, so there's no point in resisting. They don't want collisions either. The professor was going to protest again, but the Geheimrath didn't let him: Do I bore you, Jakob? You can't hide anything from me. I read your face like a book. Let's talk about the war. If the institute is the mirror of the state, it follows that Germany must be led by a negator responsible only to the highest power. You agree, Jakob, that someone has to dictate what is forbidden, lest every atom decide its own path. What will Germany become without a leader? A gas! German physicists know a few things and one is how to reason by analogy.

Another thing they know is hydrodynamics, by which our submarines have exacted a fearful toll from the English. The solution of hydrodynamics problems is in the interest of all humanity. The motion of vortex rings in a perfect fluid was discovered by Helmholtz and expropriated with inadequate citation by the English, who didn't fully understand it. The advantage lies with us. My scheme is to fire indestructible whirlpools at English U-boats. Victory may depend on my calculations, which I want you to go over, Jakob. We'll see how much you care about victory.

Waving his arms, the Geheimrath scattered papers over the laboratory floor. The professor was on hands and knees scooping them up and trying to pull some out from under the Geheimrath's boots. Will you look! Jakob be-

lieves in anti-U-boat vortices. Maybe he thinks they'll make a bronze statue of him in Berlin to stand beside Röntgen's and that he'll get a patent for his trouble. The Geheimrath added slyly: Why don't you find a solution to the equations of the world-ether for antigravity? Then we'll lift their submarines from the water and knock them out with sky mines on the way up. Look, Jakob is finding errors in my calculations! Everyone knows the kind of errors he makes. Looking up sorrowfully, the professor said: I once made a serious error in a paper. It was when I was running the institute and trying to fill duties that were new to me, and I wasn't feeling well besides. My error was pointed out in print. I was afraid that no one would take me seriously again, but I was forgiven, or the error was forgotten, or something. The Geheimrath was unmoved: You admitted to error! The trouble with you, Jakob, is that you have no convictions. Maybe you didn't make an error, but a discovery. You should have insisted on it. No wonder you've had so little success.

From behind him, the professor heard that history teaches men and nations the brutal truth: learning comes only by blows! Turning around, he saw the Privatdocent, the assistant, and the two advanced students looking down on him pitilessly. The custodian had turned up too, looking pleased with himself. Thumping the professor's head with the flat side of a rapier, the Privatdocent said: We have come for satisfaction. A student added: For the totality of your offenses, which we've borne without complaint to now. For the criticism, specialization, and individualism you've spread, which has made a chaos of culture. Professor: I teach theoretical physics, which imposes order on what had no order, on chaos. Student: For your mechanistic materialism. Professor: Theoretical physics has nothing to do with that. Privatdocent: You teach that the world-ether is a corporeal body that constitutes the world, and you say that is not materialism! Students together: Hear, hear! Professor: No, no! Privatdocent: Your world-ether doesn't exist, only your materialism. Assistant: The true world-ether can be seen by spiritual eyes alone. Student: Your teaching has left inner scars, but our idealism is intact, thanks to Schleiermacher's teachings, which answer your materialism. Student: You oppressed us with your thousands of facts, formulas, temperatures, and velocities. Professor, indignantly: It is the indispensable exactness of physics you are speaking of. Student: The cold abstractions of physics. Professor: The way to the great connections of nature. Student: We want understanding and you give us doubt. Professor: Natural science has recognized limits, beyond which lie freedom of will and the ethical force of European man. Student: You teach meaninglessness. Professor: I recognize incomprehensibility.

The professor tried to get up from his ridiculous position on the floor, but the custodian held him down. Student: You've heard, of course, that there's a revolution today. Professor: You mean the last hundred years of science and technology. Student: You admit the enemy. Take back your hundred years and give us something we can feel, something for our blood. Privatdocent: Give us reality, atoms. Assistant: Dematerialize physics, free the spirit. Geheimrath: Annihilate the English with their Manchester selfishness, shopkeeper mentality, and party strife. Student, ecstatically: Fragrance, joy, life! Professor: Reason, truth, justice . . . Student: Beauty. Professor: . . . and beauty. Student: Return to the classics—Dionysius' dreams, Ithaca. Geheimrath: Return to the classics—Lysander's victories, Sparta. Privatdocent: Return to the classics—Democritus' atoms, Abdera. Professor: Return to the classics—Archimedes' laws of equilibrium, Alexandria, Syracuse. Student, pointing his rapier at the professor's unscarred cheek: Nothing. Professor: I didn't belong to a corps. They wouldn't have me, and my father didn't approve of the Burschenschaften because of their drinking and dueling. Privatdocent, feinting toward the kneeling professor: There! Professor, wincing: I was interested in science and joined a scientific society. Geheimrath: Clearly you've always been faint-hearted. Professor: Oh, no, we drank beer after we discussed a scientific paper. We didn't wear caps like yours, but we wore beer ribbons on our watches.

The Privatdocent interrupted the professor's boring explanation: There can be no question of reconciliation. Let's settle this affair of honor at last. Wearily, the professor stood up: Our scientific society gave satisfaction, but I'm too old now. The custodian handed rapiers all around. The Privatdocent lunged. With the tip of his rapier still resting on the floor, the professor was the picture of dejection. Satisfaction now! The Privatdocent lunged again with what should have been a mortal blow. The fatal tip was a button's width from the professor's heart when the rapier went spinning in the air. The Geheimrath had reached around to parry the thrust: Jakob, you owe me your life, but I don't know why I saved it. Retrieving the rapier, the Privatdocent and the Geheimrath went on dueling in high spirits. They stood rigidly, by the rules, moving only their right arms and never so much as once blinking. When the whole party left to drink beer, the custodian stayed behind to wipe up the blood. They hadn't asked the professor to join them, which was all right. He called after them: My God, this is coarse! Something has to be done.

THE WORLD-CITY

IT HAD GROWN A LITTLE LIGHTER, BUT IT WAS NOT YET day. Moonlight shone through a break in the clouds. He looked wildly around him to see what kind of place he was in. Then the clouds closed, and his only sensation was the cold.

In no condition to go to the institute this morning, he was resigned to the study couch. Naturally he thought about his shrinking circle of friends. Sadness at the death of a friend was universal, almost impersonal. Perhaps it helped a little to see things as a physicist, but the loss of a friend was absolute. Paul Drude was a lost friend.

Jakob thought of Drude as the last classical physicist, a bittersweet characterization, more affectionate than accurate. Nearly twenty-five years had passed—could it have been that long?—since he met Drude at a meeting of the German Association of Natural Scientists and Physicians.

Jakob gave a short talk on some optical problems arising from Maxwell's electromagnetic theory of light, after which Drude introduced himself, explained that he was interested in similar problems, and promised a whole treatise on the subject soon. For today's physics, Drude said, nothing was so important as Maxwell's theory, an insight that he had only recently gained and that many Germans had yet to gain. He had taken his degree in the same year that Hertz began reporting his experimental confirmation of Maxwell's theory, which was a decisive conjunction for him; for he then began to examine his own subject, optics, from the standpoint of Maxwell's electromagnetic theory. At first he could not decide between Maxwell's theory and the older mechanical elastic-ether theory of light, but he soon recognized that Maxwell's theory was responsible for the whole recent advance in the understanding of light, electricity, and magnetism. Hertz put his finger on the reason why: Maxwell's theory had introduced an essentially new principle into physics, that of contiguous action in place of action at a distance. That principle was as important as it was difficult, and as Drude said all of this in his youthfully confident manner, Jakob felt he might have said it in the same words.

The two of them stayed away from the rest of the physics talks that afternoon and went for a walk instead. Jakob described a small, delicate research he was contemplating, and Drude described his bold plans and his stimulating correspondence with Hertz on electromagnetic optics. By the end of their walk, Jakob knew they were destined to be close friends. At dinner their talk turned to external matters, which meant that they discussed which positions might be vacant soon, which was the same as discussing, frankly, who might die. Any vacancy that opened up would start the welcome avalanche of professional moves and, among beginning physicists, dreams of a first appointment. Jakob already had his appointment and was past forty, and his best chances for a good career were behind him. But Drude had not yet turned thirty, and his chances lay ahead. He was still a Privatdocent at Göttingen and had nothing more than student fees and a small stipend to keep him going. He sent off a steady stream of publications to Althoff as he waited impatiently for his first call. Jakob's heart went out to him.

Rummaging inside boxes of correspondence, he found Drude's letters. The earliest of them told of his long frustration, his bad luck, and the ministry's blind side to his overdue claim on a position. Drude was first on the list of candidates at Hannover, but the call went to the third, and no one but Althoff could have turned it around like that. To top it, Althoff spread rumors that Drude thought he was too good for Hannover, which made Drude seem ungrateful and pretentious. Jakob turned over letters quickly until he came to one reporting that the long wait was over. Drude was moving to Leipzig to teach theoretical physics. Hertz had died earlier that same year, 1894, and Drude's great treatise on Maxwell's theory came out then. *Physics of the Ether* was a celebration of Hertz's experimental confirmation of Maxwell's theory, even if for didactic reasons it did not follow Hertz's mathematical presentation of the theory. If Hertz was to have a successor, Drude was surely the one.

Jakob regretted that Hertz did not live long enough to learn about Drude's inaugural lecture at Leipzig. He would have approved whole-heartedly. Drude might well have had Hertz's example in mind, for he argued that the experimentalist and the theorist should be one and the same physicist, not separate specialists as they increasingly tended to be. Separation made little sense to him, since theory took its direction from experiment and in return it directed future experiment. Jakob took to heart what Drude said. Unnourished by experiment, theory was a gray and lifeless thing. The same might be said of a career. Jakob had been too confined to calculation. Theory in physics! Drude couldn't have chosen a more imposing theme for his inaugural lecture.

At the end of his Leipzig stay, Drude wrote an obituary of the local physicist Hankel, which Jakob found moving more for what it revealed of Drude than of Hankel. Drude regarded Hankel as an exemplary physicist with indomitable powers of work and a lifelong undiminished zeal for discovering truth. At eighty-five and nearly blind, Hankel worked on, setting up his apparatus by feel and delivering five papers in the last three months of his life. Drude spoke of his wish that the scientific flame would radiate into the autumn of every physicist's life, as it had into Hankel's. That was only seven years before Drude's untimely end.

Drude was called from his subordinate theoretical physics position in Leipzig to head up his own physical institute at Giessen. At thirty-nine, he was on schedule. He was now appointed editor of the *Annalen der Physik*, the favored journal of German physicists. The upward curve of his success was there for all to see. It was natural that universities should try to lure him away, but he stayed in Giessen until the first university in Germany called him. That was in 1905, in some ways a remarkable year in physics.

To strict order and quick diligence / Succeeds the most beautiful prize. The poet's lines seem to have been written for Drude—and to mock Jakob! Order and diligence had decidedly not brought him what he most desired. He needed a better mind for that. He was no Helmholtz or Planck or Drude.

Or did they mock Drude as well?

Scarcely more than twelve months later, Jakob had been telephoned from Berlin. Since he had just finished his lectures, he went there directly. The commemoration was held in the great lecture hall of the physical institute. Officers of the student corporation stood by the bier as honor guards, while students and assistants from the institute laid flowers. In the name of the German Physical Society, Planck laid a wreath, and the rector of the university, the president of the Imperial Institute of Physics and Technology, a representative of the Prussian Academy of Sciences, and other leaders of science and learning all spoke of the loss to physics and of the personal qualities of the man. Only after all that had been said did Jakob fully acknowledge what had happened. Drude dead!

Still it was hard, even after these twelve years. When other outstanding young physicists died—Hertz by disease, Moseley in the war, Pierre Curie from an accident—Jakob was sad but he understood. It was the same when Boltzmann died by his own hand, since Jakob knew that he had long suffered from depression. (Boltzmann died just two months after Drude, and Jakob was asked to write a joint obituary. He declined this morbid invitation to speculate on the sickness of German physics.) To this day, he returned again and again to the act that carried Drude away. As if Drude might help him, he now looked at the photograph on the wall. He was struck by a detail, the simple rotation of Drude's shoulders to the right, of his head to the left. To anyone who hadn't known him intimately, the angle looked natural and relaxed. But to Jakob that simple twist of the head subtly expressed what the eyes unmistakably expressed: an intensity, a determination to do something important in the world . . . maybe. Jakob realized that the photographer had arranged Drude in the fashionable pose of the day, and anyway Drude was only thirty-six, nine years away from the act. There was little understanding to be had by merely looking at an old photograph. Jakob returned to Drude's letters in his lap.

He felt Drude's presence as he picked over the letters. Drude wrote that he experienced great joy every morning as he went to the institute again. That was how Jakob remembered him. Drude complained of having little time. He always took on too much and had to rush everything. That was his nature, Jakob knew. There were plenty of younger people, Drude wrote, who were at least as good as he was and who could be hired for less. He always underrated himself. Jakob knew that too.

Greetings to the new laboratory! / And let everything else be satisfactory! Despite the press of work, Drude never overlooked a colleague's happiness, here Jakob's own little laboratory, at last!

Jakob came to letters and postcards that Drude wrote on vacations. Here was Drude bicycling along the Rhine or the Mosel, skiing on the Feldberg, climbing in the Tirol.

Drude wrote that Wiener had fallen and hurt himself recently. So had Wachsmuth. Drude himself had taken a tumble in Switzerland. It seemed to Drude that every physicist had to fall.

If physicists were drawn to heights, Jakob was certain of one thing about himself. It was that he lacked the verve of his climbing colleagues. He imagined them hooking their fingers into mountain cracks and thinking nothing of dangling their feet a thousand meters over a grinding glacier.

Only once had he let himself be talked out of his native good sense. In the Swiss Alps, balancing on a narrow ledge over what was for him a fearful drop, he saw (and wondered if he had fallen) gigantic human shadows. They were surrounded by rainbows, like holy light. Later his colleagues explained that he had seen the Rigi ghost, which appeared whenever there was a wall of fog in the west and a rising sun in the east, and it was well understood by the optical laws of fog drops. Jakob resolved to keep to the gently sloping paths, where he would meet no ghosts and where he would pause to push aside leaves and twigs with his walking stick to observe a mushroom.

Jakob now held a letter from Switzerland, sadly close to the end of the pack. Drude was relaxing there before taking up his new post in Berlin. At that moment, as he looked on the beauty of the Jungfrau, there began a train of events as seemingly inexorable as if Sophocles had plotted them. As Jakob learned, Drude was reached by telegraph in Switzerland. His wife had fallen ill and had to have an operation immediately, and he hurried back to her. As she lay in the clinic in Giessen, he gathered their children and possessions and went on ahead to Berlin. With this ill-starred move, which taxed his strength and emotions, Drude's brief life in Berlin began.

Jakob tied up Drude's letters and returned them to their box. He leafed through papers in the cabinet until he found what he was looking for, an obituary of Drude—rather the manuscript of one, which he had been asked to write at the time by the local science society. He had reread Drude's publications to prepare for it, beginning with his dissertation and early writings on optics in the late 1880s, then moving ahead to his writings on Hertz's waves and then to his most profound work, his long studies of the electron theory. The newly discovered electrons promised a natural extension of Maxwell's electromagnetic theory, which joined the physics of the world-ether with the physics of matter, and Drude was naturally interested. As Jakob went over Drude's early writings, he was touched by their youthful exuberance, in a way by their defenselessness. It was easy to see where Drude got carried away with premature pictures of the molecular world, with hypothetical assemblies of electrons into ponderable atoms. (To explain the natural optical rotatory power of crystals, he assumed that electrons rotated either in right-hand or left-hand spirals, which was a rather artificial mechanism.) In time, he tempered this early boldness by his deeply rooted classical consciousness. Then fully aware of his authority and responsibility in physics, he set out every research with caution, advancing into the region of the unknown from a secure base of knowledge and not from speculation about the constitution of atoms. Jakob skimmed over his analysis of Drude's published researches until he came to the part of the obituary dealing with the end of Drude's life. Ordinarily this was the place for a few carefully worded sentiments to wind it all up. But Jakob saw that he had made it the starting point of a long discussion about—since he couldn't remember exactly what, he turned the pages looking for key words—about the whole recent development of physics!

Jakob knew in 1918 what he couldn't have known when he wrote the obituary: the years that bracketed Drude's life in Berlin, 1905 and 1906, saw the transition from classical to modern theoretical physics. In those years, Einstein sent off to Drude's journal a series of researches, the most remarkable of which contained the light-quantum hypothesis. It said that light at high frequencies behaved like molecules of an ordinary gas with its energy distributed among spatially discrete quanta, which seemed to conflict with a wealth of painstaking measurements confirming the continuous wave theory of light. Einstein singled out the reasoning in one of Drude's major studies of the electron theory to show the untenability of the classical theory of radiation. Einstein was right about the untenability, but if he was right that light quanta were implicit in Planck's original quantum theory of radiation—in the opinion of the best physicists in 1918, he was not—the equations of Maxwell's electromagnetic theory were inexact.

That was a distressing conclusion for Jakob, who counted these equations among the most secure possessions of physics. They were the foundation of his work on the world-ether and of Drude's mature work as well. But even if Einstein had gone too far with his hypothesis, by now nearly all physicists had accepted the need for discontinuous quantum processes, which Einstein had insisted on and which were incompatible with certain classical theories. Although the quantum theory had barely begun to receive attention by the time Drude died, he had thought about it. Nothing significant in physics escaped him, especially since he was editor of the journal in which Planck and Einstein published their work. Here as in all fundamental questions of physics, Drude showed his remarkably penetrating insight. In the revision of his great textbook on optics, which appeared in 1906, Drude spoke of Planck's quantum theory as the greatest recent advance. Even though it fell outside the boundaries of optics as they were usually drawn, Drude gave it emphasis; he was determined to show that his specialty was not left behind by the new developments.

In his account of Planck's quantum theory, Drude gave

the words *universal* and *absolute* a kind of telegraphic urgency. He referred to the entropy of a system of elementary radiating bodies, or resonators, as a universal function of the probability of the state and to Planck's radiation formula as a universal law. He characterized the two new constants appearing in the formula as universal, based as they were on the universal properties of the world-ether together with the universal laws of gravitation and the velocity of light. The constants yielded an absolute system of units for length, mass, time, and temperature, the fundamental magnitudes to which all physical measurements were in principle referred. Drude's choice of words revealed the kind of physicist he was, just as it revealed the kind of physics that was then emerging. There was no doubt that Drude saw Planck's theory as bearing on all parts of physics and that he saw its startling implication of atomistic energy elements, as he called them. At the place where Drude remarked that the electrodynamic significance of Planck's mysterious new constant h, the elementary quantum of action, was still to be determined and was of the greatest interest to physics, he cited Einstein's researches in 1905 and 1906 on light quanta and their relation to Planck's theory. Jakob was certain that if Drude had lived, he would have looked for the significance of h in his beloved electron theory.

The quantum theory was not the only major development in physics in 1905. That year Einstein published the special theory of relativity, and Jakob's obituary of Drude touched on it. Like quanta, relativity had to be carefully examined in its relationship to all of classical physics, which made Drude's loss to physics at just this time all the more serious. Because of the central place of the ether in classical physics, Einstein's dismissal of it as superfluous was especially troubling to Jakob. What then became of Drude's favorite expression *physics of the ether?* The ether conceptually unified all that was known about light, heat, electricity, magnetism, and much more, so that without it physics threatened to become chaos. No doubt Drude would have been less troubled, since he had long conceded that one could get on with physics by doing away with the word *ether* and speaking instead of the *physical properties of space.* But that really didn't change anything, and Jakob was sure that he and Drude were in essential agreement. In the revision of his optics textbook, Drude said that Einstein's theory was naturally no explanation of the experimental failure to detect absolute motion relative to the ether at rest. It was an elegantly simple mathematical treatment of the phenomena attending the relative motion of observer and light source. That was Jakob's view of the matter, too. The importance of Einstein's work was much clearer in 1918 than it had been at the end of 1906, and in his obituary of Drude he said little about Einstein and little more about Planck. Like Drude, he was mainly interested in the electron theory then.

Drude's editorial work didn't make his life easier. There was that difficult man Bucherer, for instance, who made demands on him. (Jakob had a reserve of sympathy for Bucherer, the theoretical physicist at Bonn, who like himself had been recently elevated to honorary ordinary professor. In some ways a sensible man, Bucherer preferred picturable, classical theories, although his were not always as clear as he believed.) Drude was advised that a paper by Bucherer was worthless and that there was little reason ever to hope for anything significant from the man. But Drude tried to help Bucherer and prevent him from making a public fool of himself—a mistake. Instead of thanking him, Bucherer played the damaged party and threatened to go over his head. Poor Drude. With his broad knowledge of physics, his even temper, and his intuitive understanding of researchers, he was a capable editor, but he edited as he did many other things from a sense of duty. It didn't help either that he had a soft heart and lacked the normal instinct for self-preservation.

The change in physics that Drude lived just long enough to glimpse did not, of course, have anything to do with what happened to him. It had only to do with the classical perimeter of his work, the original thought that had led Jakob to write in his obituary that Drude was the last classical physicist. For Jakob it was a way of fixing Drude in his own thoughts. He hadn't handed over the obituary to the editor. It had grown too long, too personal, and it had led him as it would have led his readers little closer to an understanding of Drude. Rereading it now, Jakob decided that it (like all obituaries) lacked perspective.

For some time Jakob had been reading works in the history of physics. He realized that it was a harmless pastime for physicists in their twilight years, but he supposed that he read with a questioning spirit. Actually he didn't believe most of what he read. Wary of contemporary judgments on classical physics, he went back to older histories, which were not written on the assumption that physics had burned its bridges in its recent revolutionary march. That there had been changes in some of the fundamental laws of physics since the turn of the century, he readily acknowledged, but the major gains of nineteenth-century physics had proved enduring. An architectural image came to his mind. Although the physical world picture had been shaken, nowhere had it been reduced to rubble; some new rooms had to be added and the ornamentation brought up to date, but the foundations remained firm. No more than Helmholtz might a physicist today dream of a perpetuum mobile, and where would physicists be today without the principle of entropy or the extended principle of least action or the concept of the field? It saddened Jakob to see the current scholastic tendency to judge the past from a superior present. How rare it was to come across sensible historical writing that placed the present in perspective and showed how the parts of physics had evolved together! (He used the word *evolved* even though he believed that contemporary vagueness of thought owed much to the preference of popular writers for biological metaphor over mechanical cause and effect.) Drawing out the metaphor, he described the laws and theories of physics as growing from its principles as branches grow from a trunk or as leaves from a stem. Up to a point, Jakob agreed with Mach's view that theories were merely dry leaves that fell from the tree of science after they had kept it alive for a while. Some leaves had given more life than others, of course, and Jakob thought that ether theories might not be the dry leaves that the current relativity crowd believed. And on the other hand he thought that certain theories of the atom were dry leaves indeed.

Although Jakob felt out of sorts with his times, he knew that they were among the best in his science. It wouldn't surprise him if the relativity and quantum theories were remembered as the most significant work in theoretical physics in the last ten years or so. He felt privileged to be where the work was done, there in Germany, where Einstein completed his general relativity theory and where Planck developed his quantum theory. He didn't know if these theories were specific products of the German philosophical mind, but he knew colleagues who thought they were. In any event, theoretical physics supported Germany's claim to be the cultural equal of England and France (though this was far from the thoughts of the Germans who spoke loudest on this point). The world would remember the accomplishments of Planck, Einstein, and Bohr—the quantum, relativity, and atomic theories—long after it had forgotten the Marne, Verdun, and the Somme. Why, then, didn't Jakob rejoice more than he did in this recent flight of physical theory?

Why wasn't he like Graetz, who spoke enthusiastically of the revolutionizing views of the new atomic theory, the new land? Or like Auerbach, who spoke of the revolution and the entirely new currents of thought in the physical world picture? At a special meeting last spring of the German Physical Society, Jakob had sharply felt his isolation from certain colleagues. Planck had turned sixty, and Einstein acknowledged his new decade with this meeting, at which he, Sommerfeld, Laue, and other leading physicists spoke about Planck's contributions. Papers to honor Planck had been printed and were now lying in front of Jakob, where they might prompt him in his work on his own coming talk. As he turned pages, he was struck by certain words and expressions. Twentieth century physics was increasingly the physics of quanta, he read (not of the world-ether). It owed its new face to Planck, who saw that the classical foundations of physics were inadequate to develop a theory of heat radiation. He read on. Planck introduced a *new, extraordinarily bold hypothesis* of atomistic energy elements, the consequences of which compared with the revolutionizing ones of relativity theory. He taught physicists that they must give up the assumption that underlay the previous foundations of theoretical physics, which was that only continuous processes could occur in nature. His theory kindled *revolutions* in all parts of physics, so that physicists now spoke of what went before as classical optics, classical thermodynamics, and the like. With Laue's recent discovery of X-ray diffraction, physicists thought that they had an instance of a new and fundamental fact that could be explained by classical theory, but they soon convinced themselves that it, too, demanded quanta. Etc. etc. As Jakob read on, clusters of words stood out: *far-reaching revolution such as physics has seldom experienced—courage of the few—astonishing success—in so short a time!*

Amid talk that the quantum was going to disclose the secret of the atom, there were also cautious words at the meeting. Planck responded to his birthday speakers by defining the present task of physics as the reconciliation of the quantum theory with the classical Maxwellian wave theory of light. In this connection, Jakob was reassured by the guarded attitude that most physicists took toward Einstein's light-quantum hypothesis, but it was unsettling all the same that Einstein did not give it up, since nature—and physicists, if belatedly—had come around ultimately to his other ideas.

Just last year, Einstein had developed further his quantum theory of radiation, bringing it into relationship with Bohr's atomic theory by relegating to chance the timing and direction of the elementary processes of radiation at the atomic level. The disturbing implications of Einstein's introduction of chance were underscored by Smoluchowski in his posthumous contribution to Planck's birthday meeting. Throughout physics, Smoluchowski said, the tendency was now to refer collective laws to the statistics of elementary processes and to recognize their simplicity as mere secondary consequences of the laws of probability of large numbers. Only the electron theory and the relativity and energy principles were still untouched by this tendency, but Smoluchowski anticipated that they too might be reduced to statistical regularities one day. Jakob thought that he wouldn't recognize physics—or the world it described, a play of chance—if this vision came to pass. It was, thank God, a vision for the future, and for now everyone seemed to agree that deeper causes of quantum behavior existed to be discovered and that the theory still had gaping logical holes.

Of all the physicists at the meeting, Jakob felt closest to the guest of honor. Planck remained for him a classical thinker, despite the speculative trends his theory had inspired. For years Planck had tried to assimilate the quantum into the framework of classical physics and limit the changes it required. He said at his birthday celebration that the goal of his entire scientific striving was to bring unity and simplicity to the physical world picture, to reconcile opposing insights rather than to exploit partial insights at the expense of the whole.

Planck valued the occasion most for informing the world that in these difficult war years the German Physical Society conscientiously kept up its regular personal exchanges and for informing Germans at the front that those at home continued the cultural life they were fighting to preserve. Objecting to the personal attention paid to him, Planck likened himself to an anonymous, conscientious miner who one day happens to strike an especially rich vein of gold. He was overly self-effacing, and Einstein put matters right by observing that Planck was not like the majority of scientists who worked hard but who might just as well have been engineers or businessmen or soldiers. Planck pursued his work with a mentality akin to the lover's or the worshiper's.

Jakob wondered if he himself fit into Einstein's scheme. If, as Einstein said, physicists were divided into forest trees and creepers, Jakob was in no doubt which he was. Yet he was a worshiper like Einstein and Planck, and only his grand theory of the world-ether lacked the objective significance of their theories. The work that had significance was his small studies in acoustics, optics, and electricity, which did not make much difference to the world picture.

Everyone who served the demanding goddess of science had to face the same despair in the end. As his powers of work fell off, the scientist held on as best he could, unable to imagine any other life for himself. Had Drude at least been spared the sad inevitable decline? If so, there was no consolation in this easy thought.

Jakob remembered that in his forties he first became aware that he no longer felt as happy about his work as he once had. He no longer followed the free wanderer's path, as it were, but the postman's path, which was known from beginning to end and offered no choice but to follow. He understood why memoirs of successful scientists lovingly explored the wonder garden of youthful memories and dismissed the wasteland of later memories of dutifully repeated tasks. Naturally he would not write his memoirs. No one would be interested.

Without conscious decision, Jakob moved to the sideboard where the brandy was. The burning sensation of the first swallow made him wince. He seldom touched alcohol anymore, for it made his thoughts and feelings come to him in an undisciplined way, as they were doing now. Taking a second swallow, he settled into his chair with his old diary.

Opening it, he was struck by how his handwriting slanted this way and that, as if the writing hand had been moved by passion. Maybe it had, or maybe he had just been in a hurry, as young people are. In any case, he found it hard to recognize the impetuous youth who spoke of free thought and who identified himself with earthworms, which live on cut up into many parts. He was embarrassed by these outpourings. Pages later, the diary told of an orderly plan for his days. Impatiently, he turned more pages, reading here and there about what befitted a man, a German, a Christian. There were more pages on goodness of heart. He was—how old?—sixteen perhaps. There was a teacher, a friend, a book or two that fell into his hands, just what one would expect as he began to develop scientific interests. He collected minerals, built toy windmills, read about inventions, bugs, stars, and the fascinating history of electrical discovery. How beautiful it is to think about the world as a scientist, he wrote in his diary, and he could have written that sentence today (but perhaps not the one that followed, not in the midst of this world war, about God's protective power among all peoples). Jakob came across a few troubled questions about the scientific truthfulness of statements in the bible. He hurried on until, finally, he came to it, the dedication, the calling of the young scientist Victor Jakob, who would never rest but would always work to discover the laws of nature.

He had chosen the life of the discoverer. Not an ordinary life. He closed the diary. As the brandy took effect, his thoughts traveled back to a different time, again to Drude—

Paul Drude, you laid your plans before the Prussian Academy of Sciences only days before you died. You told them of the scientific meaning of Maxwell's and Hertz's work for the physics of ether and matter. You told them of the other, technical meaning of Maxwell's and Hertz's work, wireless telegraphy. With the electron theory, you were going to develop optics, and with the guidance of Maxwell and Hertz you were going to develop wireless telegraphy. For years ahead you planned your work to avoid the danger of hurried research. It was a great joy to live as a physicist in a time like this, you said.

You weren't always sensible. Like one of those delicate receivers you designed for Hertzian waves, you selectively absorbed the currents of thought about you. They were always the best currents, but not always the best for you or any physicist to act on. Instead of settling for either pure or applied physics, you taught and did research on problems that would react directly on both. It was the age of technology and natural science, as you put it, and you weren't going to miss it. In your research, you were equally theoretical and experimental, an admirable breadth that few still tried for. You were equally researcher and teacher and your students profited, but your colleagues counted only your publications. In research you stuck with your specialty optics, but you regarded it as a discipline that sought connections with neighboring disciplines and, ultimately, connections between all parts of natural science. Then you took on the two most onerous tasks of German physics, the editorship of its principal journal and the direction of the Berlin physical institute. You were the complete physicist, and in your own eyes you never did enough.

You couldn't rest. At best you found momentary distraction in sports. Planck saw you working out as usual only two days before. Trim in your gymnastic suit with its dark sash, you looked like someone who got more than his share of satisfaction from life. When you took flight on the parallel bar or the wooden horse, you gave the impression that you could handle any problem with ease and confidence. If you had lived in another age, you might have been an explorer who united mind and body in adventure. Now discovery is done in institutes, and the body is exercised in the mountains as a cure for the mental strain of institutes. At least you had the mountains where you constantly tested your skill and endurance, where your striving was rewarded as it was not always in the human world below those heights. Below, in the Berlin physical institute, you suffered more than any of us dreamed, but you did survive that first and hardest year. The ministry had given you permission for a vacation, and you had made arrangements for assistants to take over the duties of the institute. Your equipment and clothing were already laid out for a hiking trip with Wien in the mountains. The will to live must have been finely balanced with the despair, for up to the last morning your thoughts were still with your active life. In fact, you were due at doctoral examinations, and Planck and the other faculty members were waiting for you.

After 3:30 there was no reason to wait any longer. How could this happen? everyone asked. After the fact it wasn't hard to recall signs, but at the time they hadn't seemed ominous. You hadn't slept well for weeks, and occasionally your powers of concentration failed you. Certainly your face showed the strain of your responsibilities, but every conscientious German professor is liable to collapse at the end of semester and disappear into the mountains to recuperate. The physicians talked of an attack of overexcited nerves, which left you defenseless against dangerous, compelling ideas that you had known how to control before. You didn't call out. You had friends.

Drude's teacher Voigt and friends and colleagues were at the funeral in Gotha. They came from Göttingen, Leipzig, Giessen, from the physical institutes all over Germany. Soon after the funeral, Jakob talked with Wien, who was shaken. He said he hadn't taken seriously Drude's suggestion that they exchange jobs, that Wien come to Berlin and Drude go to Würzburg. He showed Jakob a poem he had written about Drude:

> I look at the dark heavenly spaces
> And don't know if I am awake or dreaming;
> Forgotten pictures move through my soul,
> Thoughts follow, which I myself don't choose,
> Sounds that died away swell up again—
> Yet the world as always goes its way. . . .
> A victor on the field of science;
> You struggled for reality and truth
> And now are lying like Siegfried on the bier.

Jakob recognized his own thoughts in Wien's. A sinister force had defeated Drude, broken his soul.

Jakob knew that Drude had missed the gardens in Giessen and the mountains and the fir forests and the country air. He hadn't been able to breathe in Berlin!

Drude lived in the official residence within the Berlin physical institute, which meant that he was surrounded by his responsibilities day and night. Outside the institute was the noisy city, while inside it strangers jostled one another as in a train station. Only Drude's highly refined sense of duty could have wrenched him away from Giessen's peace and quiet. In exchange, Berlin offered him a single reward: I have dared it, he could say, and what courage it must have taken to walk through the entrance of the Berlin institute the day he became its master! There were the temperamental junior faculty and under them the demanding assistants and recent graduates and advanced students. There were the Privatdocenten, all impatient for advancement, and the hordes of elementary students. And there were mechanics and custodians and the rest of the staff that kept the whole intricate organization from collapsing from one minute to the next. Drude had to contend with all these people and more. He had to administer the institute and the instrument collections, deliver experimental physics lectures to huge audiences, conduct laboratory courses for beginners, pharmacists, and advanced students, direct the colloquium, and examine students from all over the university.

All of this was only his official responsibility, which Jakob suspected he didn't know the half of. Besides that, he had his editing and his work for the German Physical Society and, at the end, the Prussian Academy of Sciences. And he was permanently on call to answer all questions from ministries and faculties about physics and physicists and questions from all sides about optics. Staggering as all of this responsibility was, he had an even greater one: to do first-class research in a time of rapid advances in physics. If he failed in that, he failed as a teacher as well, and with him the university failed to represent his science. Only toward the end of his Berlin year—so great were the external demands on him—did he steal time to pick up his own work again. He could only have felt dissatisfied at best. At worst, Jakob thought, he could have become ill by worry. Of course, overwork alone could never explain the act.

Sitting in his study chair, Jakob's thoughts turned to Berlin in 1906. Everyone had hurried out after the commemorative service, and he pretty much had the institute to himself. He wandered about the corridors, sticking his head inside a laboratory now and then. He even lost his way momentarily in one of the gloomier passages. It was a giant labyrinth, this Berlin institute. It was Drude's last home.

With characteristic energy, Drude had set about to reorganize the institute. But so many letters had to be exchanged with the ministry even to replace a light bulb that it was small wonder that Drude's traces on the institute, materially speaking, were faint. He rearranged the instruments, and he posted enormous printed rolls that revealed the basic laws of physics to passersby, but even such a minor improvement as dimming lights for the lecture hall remained a promise. Superficially, every physics institute was like every other with its big lecture hall for experimental physics, its small one for theoretical physics, its elementary laboratory, and so on. But in truth every institute expressed the personalities of its present and past directors. It couldn't pass unnoticed that Drude kept his chalk on the left side of the blackboard instead of on the right where it had lain for thirty years, through the tenures of Helmholtz, Kundt, and Warburg. To Jakob it was just that placing of the chalk like any of a hundred other physical details that animated the institute with individual personality and made it a living monument to its directors. It was hard for him to accept that only days ago the corridors of the institute rang with Drude's boisterous laughter.

Outside the Berlin institute, Jakob breathed more freely. How massive the building looked from the entrance! It was hard to believe that it was not the grandest physical institute, as it had been when it was new. In power plants, use of floor space, and countless other features it

Hotel & Pension
zum
weizerischen Alpenklu
Uri
Franz & Jost Indergand
Eigentümer

was now outclassed by a good many institutes, and in sheer size it was less than Strassburg's and Göttingen's institutes and much less than Zurich's and Leipzig's. Of course, size alone did not guarantee an outstanding institute, and if proof were needed Leipzig was it. On the other hand, the old saying that the work of an institute was in inverse proportion to its size was a comforting lie of little institutes.

Jakob looked across the River Spree at the tourists watching the canal boats. He wondered if they ever noticed the reflection of a four-tiered structure nearly blocking off their horizon. If they looked up at the right time they would see hundreds of people streaming in and out of it. Their Baedeckers wouldn't help them, since the physical institute was not regarded as a cultural stop. For a moment they might even confuse the building with the royal palace a few streets away. Jakob walked ahead without looking back. He never wanted to see the institute again, for he was angry at it and at all forces today working to overwhelm the individual.

The individual! Threatened right here in his sanctuary, the German university. Jakob easily remembered back to when a beginning physicist was expected to come up with his own ideas and to decide which parts of physics were already worked over and which parts were promising. Today a beginning physicist with the necessary money simply announced himself at a physical laboratory and promptly received a research theme from the director and laboratory conveniences for dispatching it. Of course, this regimented advance of the director's reputation was at the cost of individual development—and it produced more physicists than there were jobs.

To Jakob it was an axiom that the individual was the source of progress in physics. Nothing could replace the individual researcher with his apparatus and his figures, with his hands and his imagination, struggling. Yet from experience Jakob knew that the most unlikely outcome of a life in physics was a major new discovery or idea. If the progress of physics in the last century were laid out on the railroad from Berlin to Moscow—the longest distance he had traveled by train and had an intuitive feeling for—the contribution of even a great physicist like Planck might extend from Berlin's Stettin Station to a suburb like Pankow. Jakob's own wouldn't extend further than the length of the platform.

Jakob accepted that individuals had to come together to agree on technical matters and to keep one another informed so that they didn't waste time. Physicists had their societies and colloquiums and Solvay congresses, and all of that was right and necessary. But Jakob didn't expect any fundamental understanding of nature to result from the organizing energies of physics.

Physics had grown tremendously during his lifetime. What importance this had for his career he wasn't sure. Very slight, he suspected. Except that what was given to the big was taken from the small! Jakob had seen this maxim repeatedly confirmed, which was why he felt such ambivalence about the Imperial Institute of Physics and Technology in Berlin. Without doubt it aided German industry by bringing the resources of physics to bear on its technical needs. But he was sure that it commanded resources that otherwise would go to institutes in small universities like his, which suffered chronic budgetary misery, overcrowding, and official neglect in general. The gifted physicist in his search for the laws of nature tended to go undervalued in Germany these days, with the result that Germany's world position in physics had fallen off. It was probably already too late to prevent an American monopoly of certain areas. Jakob was convinced that the root problem was the wish to centralize everything in Berlin. He was back where he started, with Berlin, with Drude.

His wife came in to ask how he was feeling and to get him out of the study for a few minutes so she could air it. When he returned, he saw that the cat had been into the cabinet, which he'd forgotten to close. Jakob got down on his knees to pick up papers scattered all over. It wasn't easy to figure out which papers belonged with which. This paper for instance . . . its formulas showed him at a glance that it was about a certain problem in the interaction of ether and matter, and its sentences looked so familiar that he could finish them in his head. Yet the writing wasn't his at all. He knew whose it was—of course, Drude's.

Wanting distance from his depressing thoughts, he decided to read himself to sleep. He might read one of Gottfried Keller's novellas, which he admired for their depiction of the education of personality, the basis of individual responsibility. He had been directed to this writer by his Strassburg physics professor Kundt, who would sit on a laboratory table, dangle his legs, run his fingers through his red beard, and tell stories about his good friend Keller. These stories would go on—and improve—after Kundt led the laboratory exodus at six o'clock to the Braun Esperance to drink Bavarian beer. Jakob didn't know if there was any writer after Keller worth reading, but he suspected not, given the absolute decline in artistic standards. Before Keller, there was Goethe, to name only the greatest of the earlier writers. Jakob loved Goethe's novellas, for their ethical idea of self-mastery and their entirely inward action and classical ideal of humanity. But this night he felt like reading in *Faust,* which was always on his nightstand.

Jakob had come with the director of the physical institute, who stayed behind at the inn to advise Althoff on candidates. Drude was there along with Planck, Einstein, Wien, and Sommerfeld, who had brought along a whole troop of junior colleagues, assistants, and students from Munich. Jakob had never been there before. Yet as if by magic he knew the scene, that wall of rock ahead, the torrents coming off it, the gorge below, the stands of shimmering birches and dark firs. Even the screech owls and plovers and hawks were old friends. The cliffs hung down like long human noses. Trees passed behind trees, and shadows of the mountains moved along the banked fog. Jakob could have sworn he was in the Harz, but the towering peaks he was looking at had a different grandeur. Planck: The light, the air! When you're up here, you leave your baggage behind, your mental baggage too. Sommerfeld: Most of all I like the new sport of skiing, but I like other sports, bicycling, swimming, and walking like this in the mountains. Up here I like to talk about physics. Everyone is here, as you can see. Einstein: This is where I turn around. Back there is a lake, where you can go sailing. Some people like water, some like mountains, I know what I like. You have to be some kind of madman to run around on mountain tops. Drude, pointing: From up there you can see everything. That's where you'll find me. Wien, clearing his throat: When I was a young physicist, I took on problems because they interested me. I didn't solve many, gave them up, didn't care. As when I climbed in these mountains, in physics I climbed quickly with no thought to the route, and I couldn't reach the top. I often remember my teacher Helmholtz, who likened himself to a mountain climber who doesn't know the way, who climbs slowly, who reverses frequently and has to find another way up, and who sees the best way to the top only too late. Like Helmholtz, I no longer expect to find the royal road at once. I don't hurry but spend most of my time choosing problems that I and my students can solve. It pays off in physics, I've learned, and it pays off in the mountains too. Jakob: That crashing water sounds violent, but the valley wouldn't be green without it. Contradictions don't seem so troubling up here, not even the ones between classical and modern physics.

Soon Einstein was out of sight, and Sommerfeld's people had stopped to talk about the atom. Planck, Drude, Wien, and the guide were in full stride, and Jakob couldn't dawdle if he didn't want to get left behind. They hadn't gone far before they met Abbe, who was there from Jena and hadn't time to stop and talk. His wife came up behind and explained: He replaces one exertion with another, physics with mountain touring. He never knows a real rest. Jakob's party continued climbing. No one said a word or stopped for hours. Too tired to go on, finally, Jakob stayed behind and watched the others climb low summits until they reached a wall that was high and steep. There they separated. Jakob could see Planck, Wien, and the guide start down a trail that ran through dense dwarf-pine forests. Drude started up the vertical cliff face, which looked dangerous and was clearly no climb for one man. After a while, Drude was unable to move—either up or down—and yet he didn't call for help. Jakob ran until his lungs ached. If only the rocks and trees would stand still! Running, arms outstretched to break Drude's fall—too late, he watched the precipice shake until Drude plunged to the bottom. NATURE'S HOSTILE FORCES *ran through his mind, and since the object of physics was to understand and control them, why had Drude pitted his strength against them? Jakob scrambled over the rubble to find Drude lying on his back, his head on a rock. He bent down on his knees, but he was almost afraid to look. What astonished him was the gun lying in Drude's outstretched hand. Drude smiled: Don't I make the perfect picture of a ruined gambler?*

THE WORLD

THE MOON SHONE MOMENTARILY. HE THOUGHT HE SAW a branch sticking out. He reached as far as he could in the general direction and felt something soft, a small dead animal perhaps. He couldn't see a thing now.

Maxwell! The cat didn't respond, and its limbs looked unnatural, stiff. As Jakob reached out to touch it, his hand started to shake out of control. In time he would have to take care of the poor dead cat.

He no longer felt up to going to the institute. From his couch he looked out the window and watched the sky pass through every shade of gray. How many times he had watched it, he had no idea. He turned over old letters and manuscripts to pass the time.

This morning his wife told him about running into the baker's son back from the front. He alarmed her by his discouraging account of the war. Something else: at the baker's she ran into the wife of the chemistry professor, who gave her two tickets to the theater for tonight. Since her husband was working late to finish an assignment for the army, they couldn't use the tickets. The play was *Antigone.* If they felt up to going.

When he was young, Jakob wanted to see Greece but hadn't got farther than Italy. His first trip with his parents was by coach over the Brenner to San Remo, where they rented a villa and an orange tree, which bore fruit of great sweetness and softness. Their garden ran down to the beach, past wild violets, hyacinths, tulips, and red anemones. In the mornings there was a light movement of air from land to sea and at night a reverse movement, and the air was always heavy with the smell of orange blossoms. It was seldom over thirty degrees or under ten at night, unlike the sultry summer days and freezing nights in Germany. The family went on to Rome, drawn across the mountains by twelve pair of white oxen. (Jakob remembered his boyhood wonder.) There they lived like Romans, going to a cafe in the morning, a trattoria at midday, and back to their hotel at night for wine, bread, and salami. They made the rounds of museums, churches, and ruins, while his father lectured him on glory and his mother on beauty. Rome was then still as Goethe had described it, *the* city.

Years later he returned with his parents to San Remo, which had changed for the worse. Progress had appeared as a road from Genoa, so that the inevitable rows of elegant villas and hotels had replaced olive groves, flowers had disappeared behind walls, and doors had to be locked at night. A gas factory stood in their old garden and a gas tank on the beach. They went again to Rome, which he found spoiled too, an eyesore of bad new buildings. Jakob shook his head as he remembered all this change.

He saw his first Greek temple in Italy. He was speechless before its power and beauty and the absolute order of the place. At that time there were few houses, mostly open fields of myrtle and in the distance the sea and the mountains. Jakob hadn't been back there, but he was certain that its holy solitude had been spoiled by smoke, noise, and the rest of arriving culture. He wouldn't like to see it now.

His wife unwrapped a modern French novel, which the bookseller was glad to get rid of since French depravity had helped bring the war on. It was not something Jakob would ever read, and not wanting to let go of his memories he asked her if she remembered the orange trees on their trip together to the South, to Greece. Yes, and she remembered the ancient olive trees and the eucalyptus and cypresses. And roses, roses everywhere, which bloomed the year round. Then he remembered ruins strewn over the fields of flowers, open theaters in the rocky wastes, and the absolutely still sea. Then together they remembered their climb from Olympia through oak and pine forests, past fields of wine grapes, toward the snowy heights of the Arcadian world. They climbed in chilling rain, up steep, stony paths, falling with their horses, losing their way in the dark, then finding it again by moonlight, at last reaching the beautiful temple preserved in its high wilderness solitude. The hard climb had made Jakob think of the ceaseless upward striving of those Greeks.

Jakob could still recite long passages from *Antigone* in Greek. For this gift he owed nothing to his gymnasium, which taught Sophocles for syntax and irregular verbs. It was from his father that Jakob learned to read Sophocles fully, for his nobility of theme, which prepared him for the greatest moment any classically educated person could know: to stand on the Acropolis and look out across the land from which all greatness came. Because an international archeological congress was meeting in Athens, Jakob saw a Greek production of *Antigone* in the Stadium, the Acropolis in full view. If there was a flaw in that remembered picture, it was that Athens had grown so large that it sprawled all over the landscape. There was another flaw: about all he heard spoken was German, and it was hard to find a Greek among the mass of tourists with their pale northern complexions.

Yet he had never been so happy as on that visit to Greece, there in human contact with the land. His happiness had another, more personal cause, his wife, who was now removing her hat, looking toward him. He had nearly turned his proposal into a joke in the middle of it, since—at his age!—he had no right to expect her to say yes.

Jakob had been in the habit of spending an afternoon a week with the classical philologist at his university. Now Jakob believed fervently in the educational value of the classics, but not that classics should dominate general education at the expense of physics and other natural sciences. The philologist argued that the educational value of the natural sciences was unclear, since they did nothing for the moral development of the individual. In reply, Jakob wondered what the classics did for the moral development of philologists. He had taken an examination in which the philologist asked him to discuss the laws of Athens in the year 394! Ever since then, he had had bad dreams about philologists, who loved to humiliate physicists, having lost all sense of balance and humor.

In fact, Jakob and his philologist colleague had the same cultural complaints, by and large, and the same affection for Greece, and they got on perfectly well. They were often joined by the philologist's daughter Helene, whose knowledge of classical literature was wide. Of course she knew nothing of the exact sciences, and as she took an interest in Jakob's stories about Archimedes, they began meeting separately. It was only natural that on their honeymoon in Greece they should also visit Syracuse in Sicily, where Archimedes carried out his studies in mathematics and physics and where he invented weapons for the defense of Syracuse against Rome.

To Jakob's annoyance, his historian colleagues were constantly comparing Germany to Rome and England to Carthage: as Rome won that war, Germany would this. But those historians didn't remember or, Jakob suspected, didn't care that the Romans murdered Archimedes at his studies during their sack of Carthaginian Syracuse.

Before Jakob and Helene left Greece, they visited the battlefield at Marathon, but they weren't much interested in war then and did it only from a sense of duty. They took a Lloyd steamer to Constantinople and had to be guided by a Turkish torpedo boat through mine fields in the Dardanelles, a foreshadowing. They looked at mosques in Constantinople, but what remained in Jakob's memory was the thousands of sick, homeless dogs in the street.

Following his wife out of the study, he went to the dining room to read the weekend newspaper. He began with the back pages. An ad from a Berlin firm offered insurance against damage and bodily harm from air attacks, with favorable premiums and conditions, everyone welcome. They were on their toes in Berlin. Everyone would soon be in danger and in need of insurance: yesterday Frankfurt was bombed again, tomorrow it might be almost anywhere. The armor plate that planes now carried showed the determination of the nations to convert the air into a battlefield, just as the depths of the sea had been converted into a battlefield by the U-boat.

His eye ran over ads for preschools and patent cold medicines and, at the bottom of the page, ads for marriage partners, mostly from women, some certainly having lost husbands in the war. A woman with an elegant figure, a lover of nature, in her early forties, was seeking a man not under fifty, preferably an engineer. That sounded sensible to Jakob. Helene said it sounded unlikely, too nice, too normal. She knew what it was like. The men at the front had their bordellos, while their women at home worked in factories and took lovers. Widows, yes mothers, were driven to prostitution. Quick marriages, divorces, births outside marriage, abortions, homosexuality were increasing at a frightening pace. This war was destroying morality. Things were coming apart everywhere, Jakob agreed, and he brought up the cheating and the stealing that everyone was talking about. Without moral bonds, there was nothing left but chaos, he concluded.

He read aloud an ad for artificial leather made of paper, since Helene needed a new coat. She said she was disgusted with soap made mainly of sand and tired of artificial bouillon cubes, artificial marmalade, honey, lemonade, and almost everything else in the kitchen. The world was becoming artificial, he observed. The concept of the natural was becoming outdated. There were war shortages, she reminded him, and there was nothing to be done about it. She poured his coffee. He tasted its acorn flavor and decided it was scarcely drinkable, while with his pencil he circled an offer of 50,000 marks for an artificial cork for champagne flasks, making a mental note to look in the low-temperature laboratory for suitable materials.

Turning to the inside of the newspaper, Jakob began reading aloud a serious essay about synthetic industries, which were going to free Germany from dependence on England and America. Although everyone now was sick of wearing paper clothes, after the war the German chemical industry would make appealing substitutes from wood for wool and cotton. For this purpose Germany's far-sighted Eastern politics would assure the necessary wood supplies from Russia and Finland. From nitrogen fixation, synthetic fertilizers would enhance harvests, and the same far-sighted politics would secure lands in the East on which to spread the fertilizers. Pessimists on Germany's world-political future had overlooked the power of German science and industry . . . His wife didn't seem interested. He turned to the news and editorials.

He read aloud reports from Baku, Japan, Palestine, Rumania, Russia, and the state of Brooklyn (he knew better), where troops were deserting. It was a *world* war, as they had called it from the start. Germany was bordered by so many—mainly hostile or indifferent—countries that it took fingers on both hands to count them, and it was bordered by the sea, which England ruled. Yes, it's all true, Helene said and went to the kitchen to check on the meal. Germany, he read to himself, had only its own power to rely on and its solemn acceptance of the burden of responsibility. For in these fateful hours, the history of Germany and of Europe rested on the German people. He was not unmoved by this appeal, but he knew that more than willpower was needed. He turned from the editorial to a communiqué from Ludendorff's headquarters, the optimistic wording of which didn't obscure the picture he formed. It was evident that the German armies were retreating toward Germany's frontiers, as they had been since midsummer. As the three

Frankfurter Zeitung
und Handelsblatt.
Preis 10 Pfg.
Sonntag 4 Uhr nachmittags
September 1918
Der feindliche Ansturm gegen die Westfront.
Erfolge der deutschen Verteidigung.
Das bulgarische Angebot.
Karte
Schmalzbezugskarte

world powers of the West moved forward in decisive battles, their gains were extraordinarily bloody. He didn't doubt this, nor did he doubt the unmentioned bloody losses on the German side. During the meal he discussed the war. Flanders, Cambrai, Chemin des Dames, Champagne, Argonne, and all the other battlefields of the West had come to life. From the coast to Verdun, the greatest battle of the world war was waging. How was it possible for armies to keep fighting like this, he asked. In Russia, he read, people were driven from their homes by the thousands, imprisoned and executed, as a whole class was being annihilated for political advantage. He read about a flu epidemic, about starving people . . . What was going to happen? To Germany? To the world?

Helene said he spent these days too much with his own thoughts. She missed the cultural life of cities, though she appreciated his reservations about those places. She read a Munich newspaper, which was relatively free of the usual German parochialism. Gogol's *Taras Bulba* was serialized in it, as if Russia hadn't been Germany's recent hated enemy. Shakespeare's *Measure for Measure* was put on in Munich as was a new German play *Burghers of Calais*, which evidently portrayed the English and French sympathetically. No fewer than eight theaters in Munich—she counted the ads—were playing. In the coming week, the Munich Hoftheater was putting on three classical plays by Schiller and Lessing and four operas. Next season, she read, Munich was putting on 250 concerts. Wilhelm Backhaus, Bruno Walter . . . she read off the names. There was a show at a new art gallery, but Jakob wouldn't know the artist and probably wouldn't like his work. She hardly bothered to mention the new movies with double features. Despite the hardships and the kitsch, culture could be found in Germany. Munich was ample proof to the world of that, as their own small town decidedly was not. Their last time out had been to an operetta, which had amusing scenes from city life, but the endless patriotic songs and cute children bored them. Couldn't the promoters raise war loans and the level of local culture at the same time? They got up from the table. They were going to the theater.

Could Germany be losing a world war right now as the theaters and coffeehouses were filling? Jakob had only to ask the question to know the answer. Yet . . . They hadn't realized the play was not to be Sophocles' until they got to the theater. The young man sitting beside Jakob turned out to be a fan of the playwright, Hasenclever, an Expressionist. Jakob had heard about this new breed of writer who went in for screams, monologues, and monosyllabic world views. If it couldn't be Sophocles, he wondered, why couldn't it be Goethe, whose eloquent, classical acceptance of nature, history, and fate moved him so deeply? By contrast, the new playwrights despised the laws of nature and the tranquility that came from willing submission to them. The dream, the nightmare, the hallucination, the apocalypse were their world, as Jakob had gathered from reading the feuilletons. His neighbor was talking excitedly, so he had a hard time following: . . . has overturned the laws of drama . . . has revealed a higher future for all humanity . . . The vision of today's youth has raised aesthetics and ethics to a higher unity. It wasn't too much to call their style revolutionary, which left the false progressives of the last literary generation shaking their heads. Today's revolutionaries were creating a direction, a movement, a literary epoch, and were exposing the old Naturalists, Realists, and Symbolists for their merely decorative and faithless style . . . Jakob's attention wandered, and he wondered if the play was ever going to begin. His neighbor persisted: The new playwright is the new man who leaves behind his isolating individualism, merges with humanity. He is the messiah who sacrifices himself for the modern world. The modern world is dematerialized activity, charged with spirit and love, the world of revolutionizing modern science, as radioactivity has proved, etc. etc.

Thank God, the play began. It was actually a reading, since there weren't enough actors left in town for a performance. It was Sophocles' story up to a point: Antigone buried her slain brother and was killed by King Creon, who had forbidden it. But it bore a new message, an appeal for Christian love of humanity. Even less Greek than its religion was its politics, for tyrannical Creon spoke just like William II. Moreover, not only did Creon lose his crown, but at the close of the play the poor masses rose, led by a Man of the People. The censors probably thought it was all in Sophocles. Little classical learning was to be expected these days, as Jakob well knew. He left the play disheartened and complained to his wife on the way home that the playwright had made a war soup of ethical commonplaces from Christianity, pacifism, and socialism. Yet he had to agree with her that the play had a certain power, hard to put his finger on. It had grown out of the needs of the war, and no one could deny that its call to humanity for healing reconciliation was needed. Love ennobled everyone, even the bearers of swords. That was Antigone's faith. Jakob had grasped that much.

The play troubled him. Like other modern plays, it portrayed characters as abstractions, as the Man of the People and so on. To him it was a distasteful way of thinking about the individual. Yet he realized that it had become a part of his own way of thinking. In his private thoughts he had become a type and a nostalgic one at that, the Classical Physicist.

If Jakob regarded himself as the Classical Physicist, that was because his waking life was increasingly a meditation on lost classical virtues. In the slaughterhouse of Europe, he asked, who still honored moderation, restraint, balance, order, clarity, and tradition? Or simplicity, proportion, and unaffected speech? Or universals, absolutes of conduct, and eternally valid standards? To what end had European youth been drilled in the classics, in the careful thought of the languages and literatures of antiquity? Had their training been inappropriate for an age of wireless telegraphy and steel airplanes? With questions like these, the Classical Physicist made little headway, and sometimes he suspected that answers were going to be determined by the war and not by educators like himself. Whatever was classical in the European spirit might not survive, and no one knew what would replace it. Would the example of classical physics—a product of the threatened European spirit—be lost on scientists to come? As he reflected on classical physics, it appeared to him to be not so much one particular understanding of nature as a set of virtues, a quality of thought and character. The Classical Physicist knew it might be fanciful, but he saw a bond between physics and the classics. Classicists penetrated to the spirit of ancient thought and life, and physicists penetrated to the essence of nature. Classics was premised on the similarity of ancient minds and modern minds, and physics was premised on the similarity of mental and natural processes. It was a bit simplistic, he realized.

Like the Classical Physicist himself, physics was becoming increasingly abstract. Physicists used to seek picturable mechanisms for understanding the world, but now many of them had pretty well given it up. Their immaterial world-ether was not a mechanism so much as a set of mathematical relations, and if they substituted empty space and energy for the world-ether they were left with an even greater abstraction. The *cold gray cave of abstraction* came to Jakob's mind.

Jakob believed that physics in its present state couldn't sacrifice the world-ether without sacrificing its goal of intelligibility. Properly formulated, its theory would retain the equivalence of mass and energy from special relativity and incorporate the two or three triumphs of the gravitational theory of general relativity and the several more triumphs of the quantum theory of atoms. It was easier said than done. He would not be the one to formulate the true theory. His ongoing, if fitful, work on the world-ether (which had begun as a bright promise of a call to a chair if not of scientific immortality) was now hardly more than an indulgence, a mere habit.

As he should have known, he had overdone it again. After his night out at the theater, he fell back onto the couch with every sign of staying there. To pass the time, he flipped through his diary from 1870, that momentous year. He had recorded the gathering political clouds, France's clumsy efforts to humble Prussia, the Ems dispatch, France's declaration of war, and through it all the buildup of tension that prevented him from giving full attention to his lectures. His physics professor didn't help either by his political asides. In the acoustics lecture he made a patriotic pun, which Jakob must have admired since here it was in his diary. In Paris the acoustical tone—*Stimmung*—was lower than it was in Berlin. That was it, and if it didn't seem brilliant to Jakob on rereading it now, it was only a measure of his remove from those events. His diary noted that the students gave the professor their jubilant approval, and there was no reason to believe that he hadn't joined them.

On the day of mobilization Jakob didn't hear any lectures. With his comrades on the way to the front, he was outwardly high-spirited, but in his diary he recorded the seriousness of the moment. They understood the historical antagonism between France and Germany, and they didn't feel racial hatred toward the French or their own Teutonic superiority. For a stretch, the diary reported marches in the field and other banal facts of military life and then, at last, his first battle. They were told that Germany had won it hands down and that, in fact, Napoleon had sent his dagger to the king. The report was cruelly false, but it had a fortunate outcome for Jakob. It was the occasion for his meeting Voigt, a serious young physics student from Leipzig. Voigt was in the Saxon army and was disgusted with the continuing particularism of his countrymen, which was out of keeping with the idealism he felt. On hearing the premature news of victory, his army hailed the king—the *Saxon* king!—and sang the "Beautiful Blue Danube"! They didn't say a word about the Prussians, Bavarians, Hessians, Württembergians, and other Germans who had contributed to the victory, which was the whole point of it. When Voigt heard "A Mighty Fortress Is Our God" from Jakob's bivouac, he came over and they fell into conversation easily. Voigt had decided on a career in physics rather than in music, since a musician had to be absolutely first rate and a physicist could get along on less. After the war Voigt went to Königsberg to study physics. Jakob didn't follow him because he had heard that experimental facilities there were poor and that not much physics was going on.

In recent years, Jakob had often discussed the current state of theoretical physics with Voigt, with whom he saw eye to eye on a number of things. They acknowledged that some principles of mechanics had been found to lack general validity and that electrodynamics needed some modification. But what impressed them was that the recent inexhaustible criticism of physical theory had not shaken its structure. They also agreed that the assumptions the atomic theorists were making these days had only the roughest approximation to reality and posed insuperable mathematical difficulties.

Jakob quickly turned the pages of his diary containing more details of battles. Here and there he paused. He read that with cries of *Vive l'empereur* that could be heard over the thunder of battle, a dense line of infantry in red and white pants bore down on them and forced them back. But their officers raised their sabers, sprang forward, shouted hurrahs, and moved up the hill again. Immediately, their captain fell, drilled through by more bullets than were needed. They struggled on all the same. It was such steep going that Jakob had to pull himself over boulders with his free hand . . .

There was a gap, followed by pages written in a field hospital and more pages in a Berlin hospital. Then the war was over, and there he was in Berlin. Dirty water ran a meter deep in the gutters of that dismal town. But he was there, and Helmholtz was the new director of the Berlin physical institute. So he stayed on.

He overheard raised voices in the corridor, Helmholtz' among them. How did it happen, Helmholtz demanded, that in his institute he and his family were not assigned windows for viewing the entry of the victorious troops? He was going to reassign the windows, and he would allow no one into rooms containing instruments who didn't know their value. As he passed by Jakob, Helmholtz was explaining that he had to assert his rights in the institute from the start.

Jakob did not have space in the institute, so he stood in the shade of the linden trees to watch. Before him, the Brandenburg Gate was decked out with flags, wreaths, and a great victory banner. The honor battalion bearing captured French field emblems marched by, an iron cross on every breast. Moltke, Bismarck, and Roon rode in, and then after a tense wait the Emperor sprang through the Gate on a beautiful stallion. He received the national hymn and gun salutes. From platforms, rooftops, and windows, people cried out their gratitude. It was a joyful moment for Jakob and at the same time a reflective one. What he had done and seen in those few months of war had changed him. The bitter conclusion of a soldiers' song came back at this rather inappropriate moment:

> And when the war is over, where do we turn,
> Our health lost, our strength gone?
> Again, you'll be called a bird without a nest,
> Brother, take the beggar's staff, you've been a soldier.

Many of his comrades were coming home with an empty sleeve or trouser leg. Jakob had been lucky with his wounds.

In the institute, Helmholtz talked about the war. What would become of that unhappy nation France, he asked. With its mad vanity, its powerless hate, France had suffered a terrible, crushing justice. Jakob agreed with this observation by Germany's first physicist.

The physical institute was still in the left wing of the university building. To one side of the lecture hall was the collection room, where Jakob worked. In the corner of the room were steps that Helmholtz came down daily to observe Jakob's experiment, but never to give a word of opinion or advice. From time to time Helmholtz talked to him about physics, and when he did he began with the simplest things and ended up in spheres that left Jakob feeling stupid. Years later he understood some of the things Helmholtz had said to him when they were in daily contact. How privileged he had been to be near Helmholtz he realized even better now than then. He had known the most noble conscience in physics, which more than made up for the cramped and miserable condition of the institute. (Before he left he saw colossal steam engines ram great posts into the swamp floor for the new physical institute.)

Jakob had been in the audience when the president of the local science society gave the memorial address on Helmholtz. (Really Jakob should have been asked to give it, since he had known Helmholtz.) The president described Helmholtz' ceaseless struggle to bring unity to physics and, in the final phase of his career as president of the Imperial Institute of Physics and Technology, to bring physics to the service of German industry. His address gathered power with his recitation of three emperors who bestowed titles on Helmholtz, quoting the most recent, William II, who complimented Helmholtz for remaining aloof from all parties and politics and for holding to the ideals of science, fatherland, and king. Then everyone shouted *hoch, hoch, hoch!*

That is, nearly everyone did. Jakob wasn't paying much attention, for he was thinking about Helmholtz' quiet inward strength, the source of his intellectual greatness. Helmholtz described himself when he said that scientists with an inner drive to knowledge acquire a higher understanding of their relation to humanity. They experience the whole world of thought as a developing entity, which is infinite in comparison with the brief life of a scientist. In his maturity, Helmholtz passed beyond all egotistical drive, bound by love to an everlasting object, which sanctified his labors as a scientist . . . That was the true meaning of Helmholtz' life in science for those like Jakob who came right after him. Jakob remembered that Helmholtz expressed these beautiful thoughts at his seventieth birthday, which was Jakob's birthday coming up. Helmholtz was right. The first seventy years are the best.

Jakob studied not only in Berlin but in Germany's newly won Elsass. When he arrived, Strassburg was crawling with bugs and vermin, but the Germans soon cleaned all that up. The physics professor Kundt along with other young professors at Strassburg were determined to bring honor to Germany's new university. From Kundt, Jakob learned to measure nature precisely and (unfortunately) to regard higher mathematics as a waste of time. Jakob made a nice measurement of the velocity of pure sound waves, working happily with primitive means and with teeth, toes, and fingers. Early in the world war, Jakob spent a week in Belgium where he had a whole army troop at his disposal to measure sound. He failed despite all the free help, for wind made the sound of enemy guns erratic.

In 1914 Jakob was in Berlin again to follow up a letter to Planck about a possible solution to the difficulties that the quantum theory posed for the physics of the world-ether. Planck had returned a thoughtful criticism, pointing out that Jakob's work belonged to the class of theories that sought to order all phenomena under a single principle. The definitive solution to this problem would never be found, he said, but progress toward it was always possible and valuable. With regard to Jakob's hope that his theory pointed a way to understanding the quantum behavior of the interior of atoms, Planck was entirely negative. Of course there could be no question yet of putting the theory to test, since Jakob had only managed to apply it to the simplest cases, a mass point moving in a static gravitational field and the like. For the principle behind Jakob's theory, however, Planck showed a guarded sympathy, and so Jakob had come to Berlin to continue their discussion.

Jakob enjoyed visiting Planck in his own institute for theoretical physics, by no means the typical Berlin institute. Its endowment was so small that any experimental work was out of the question, and it had only one assistant, who corrected papers. It was just about the only place in Berlin where Jakob didn't feel overwhelmed.

Planck did research at home, which was by the forest and lake to the southwest of the city, in the Grunewald. His house was set among the blue-green pines, behind ferns, primroses, and crowfeet. Professors and their families on evening strolls through the colony often heard music coming from his house, from the balcony door on the upper floor, the cluster of windows beside it, and the bay and arched windows on the ground floor. They knew they were hearing Planck at the piano, since they had often watched him playing while guiding a choir of daughters, neighbors, and physicists through a song by Haydn or Brahms. There in that house, Jakob heard Planck accompany Einstein on the violin. The depth of feeling that those two theoretical physicists imparted to their music impressed him. It was, in fact, while he was participating in the cultural life of Planck's house that Europe went to war again.

The mobilization of Austria-Hungary prompted daily

demonstrations in front of Jakob's hotel in the heart of Berlin. In the pubs they shouted Hail to the Triple Alliance! Shame to the Balkans! Newspapers were full of talk of guns, battleships, ultimatums, and honor, of the coming world war, the twelfth hour, and the greatest crisis of the Reich, culture, and science. Through it all, Jakob expected that the governments would come to their senses and back down, as they had before. On the last evening of the peace he milled with the crowd in front of the palace of the chancellor. From a middle window, in booming voice the chancellor spoke of Bismarck, Moltke, William I, the Reich, and the last drop of blood. The next day was a brilliant Saturday, bringing every Berliner into the streets to cheer the castle guards and clean out grocery stores. It was war, and already there was bitter grumbling about the increased price of potatoes. It all made Jakob tired, and he looked for a place to sit. But the coffeehouses and restaurants were full, so he had to push through the mass of people surging with the marching columns to get back to his hotel. He hardly had the strength to untie his boots. His feelings were not elation and just possibly the opposite, for he sensed that a great European tragedy was beginning. The throng under his window reached a peak of excitement in singing "March of the Entrance into Paris." He fell asleep in the middle of it.

Soon after the war began, he received a letter from a respected colleague in theoretical physics whom he had always thought of as reasonable. The letter contained hatred such as he had never heard, though he would soon hear more and worse. The point it made was that every German lived for the moment when the English along with their fleet and their cities were wiped off the face of the earth. Soon after, Jakob learned about Lenard's call to Germans to erect a continental blockade in thought and action— especially in physics— against the English until such time as they were completely broken or annihilated. Germany, rest assured, could carry civilization by itself with no help from the English. Then he learned that Wiener was lecturing to his physics pupils about opposing world conceptions for which the war was fought, the money morality and egotism of the English and the social sensitivity and altruism of the Germans. These outbursts and others like them made Jakob acutely aware that the national mood touched physics as it did everything else and that it acted on the reasoning mind, his own included, as inexorably as gravity acts on stones.

He didn't put much trust in the familiar characterization of the English as utilitarian, materialistic, piratical tradesmen. In his own science he regarded Newton as incomparable and Maxwell as next after him. The present great English physicists he also looked up to, and he cited their works scrupulously. He spoke their language more or less and went to their meetings when he had a chance. He even gave a talk to them just before the war in which he said that by appearance, language, and history, the Germans were first cousins to the English, really closer than the French and Italians. Humanity was indebted to both peoples, to the Germans for returning to the greatness of Greece, to the English for Shakespeare; and natural science was indebted to the Germans for Helmholtz, to the English for Darwin. The English liked his talk, it seemed. He had always overlooked small affronts, as when an English physicist complimented Hertz by saying he wasn't an ordinary German but one of the highest type. That was only to be expected from a healthy rivalry between the two best scientific peoples.

But in 1914 when England declared war, he felt betrayed, and when the Oxford professors declared war against Germany all over again, he was not sorry to see them answered by the Cultural Alliance of German Scholars and Artists. He didn't sign their petition, but Planck did, so his science was represented. English scientists and scholars said such unfriendly things about the Germans that Jakob at first even sympathized with the manifesto signed by Sommerfeld, Wien, Wiener, and other German physicists. It argued that in the past English physics had enjoyed an unjustified influence in Germany and that it must not be allowed to continue. On reflection, Jakob disapproved of the manifesto, not for its content but for its pointlessness, even pettiness. While this war was going on, no publisher would have the courage to publish an English work on physics, and no physicist would publish in an English journal. It was hardly worthy of German physicists to take advantage of the war to displace their English colleagues, and in any case their manifesto would only be seen abroad as an act of reprisal for imagined past injustices. Jakob noticed that the manifesto was not signed by Planck, whose judgment he usually found congenial with his own.

At the beginning of the war, Jakob exchanged a number of letters with a friendly English colleague through the American consulate or through Norway. But he felt the strain, and when the Englishman wrote that he had just enjoyed watching a zeppelin burn, he knew it was time to break off the correspondence. Foreign physicists who had been working in England and Germany at the outbreak of the war represented the international spirit of science one day, and the next day they were interned as hostile aliens. Pringsheim was interned in Liverpool's German concentration camp, where he lectured on physics and was desperate to get physics papers. The English physicist Chadwick was interned near Berlin, where in the camp university he lectured on radioactivity despite the lack of demonstration apparatus. Chadwick's co-worker Geiger said that Chadwick was atoning for the sins of the English and that he was much better off than Germans were over there, and if Geiger's remark sounded unfeeling it only reflected the exclusive national feeling that gripped almost everyone. Early on Jakob had arrived at an understanding like that of Planck and Warburg, which was that neither Germany nor England had a unique civilizing mission in the world, that all countries contributed. Germany was fighting to defend itself, but Jakob had come to see that in a certain sense England and its allies were also defending themselves.

Cigars were good for quieting hunger pangs. Early in the war, genuine tobacco was easy to get, but now Jakob had to trade on a long friendship to get decent cigars. He was under orders from the doctor to rest and not smoke, and his wife was there to enforce them. So it was with stealth that he put on his coat and let himself out of the apartment and made his way, weakly, to the house of his friend the tobacconist. He came out again with a lighted cigar in hand and wandered over to the town square. There he sat down on a bench. Soon smoke was curling about the knees of the statue of Bismarck standing above him.

How different it had looked four years ago! The Emperor's declaration of war was posted at every corner of the town square, flags were out, bands were playing, everyone was throwing flowers on the marching soldiers, and boys from the gymnasium were standing in line to join them. In uniform, a student from his class stopped him on the street to say that the war brought out generous qualities in everyone. Even professors seemed generous for once. Professors had always found fault, condemned Germans for their irreligion, immorality, materialism, and values, and deplored their parties, confessions, classes, and eternal quarreling. Now the professors were loved as they praised the people and even for once the Emperor. It was all a wonderful dream, a beautiful vacation: professors and students, officers and foot soldiers, factory owners and workers, liberals and conservatives, Catholics and Protestants, Saxons and Prussians were all united behind Germany's just cause. Here was a great opportunity for talks, which the professors didn't overlook. It came natural to them to speak of the eternal meaning of dying for the fatherland and, this side of dying, of the moral purification of war. They spoke of the individual who lost all egotistical traces in selfless dedication to the whole.

They didn't ask why they were ready to kill the man who spoke another language, why they readily abandoned their high ethical behavior as members of cultured nations. Later some of them did ask that question. Too much had happened since 1914 for them still to take an uncomplicated view of the meaning of the war for Germany and the world. Just the other day he heard from a young mathematical physicist who had been wounded and was out of action. He said that at first he had been enthusiastic for the war, but now he only looked to the day when he could return to the Germany of Einstein and Hilbert. If that was an indication, Germany had survived the war, whatever the military outcome. For preserving cultural values at home, the older scientists and scholars would seem to deserve praise.

But they were blamed! The young mathematical physicist was not alone in attributing the painfulness of the war in part to lazy, unthinking intellectuals, who along with generals and politicians led the country to the edge of the abyss. Lately, Jakob had heard this sort of criticism from several quarters. Professors, the critics said, had spread hate through their teaching, had subjected youth to their murderous fantasies, and had justified all manner of force and brutality. Jakob didn't think that physicists had done much to spread hate. He didn't think they had done much to mitigate it either.

The philosophy professor from the university sat down on the bench beside Jakob and brought out his cigar. He asked Jakob if he should say anything about Einstein's relativity theory in his lectures on the theory of knowledge. Could a general relativity of thought and value be inferred from it? If philosophers could believe a statement and its opposite at the same time, wouldn't they be back two thousand years with the Greek Sophists? Where did that leave truth? Jakob thought that truth was always in danger this way, but not because of relativity theory, which had to do with the uniformity of natural laws to all observers, not with relativity of judgments. Physical knowledge was objective; the consequences of natural laws agreed with everyone's experience. The philosopher went off reassured. After a while, the professor of political science came by with his cigar. He pulled a slip from his coat pocket with a formula describing the economy of a state without taxes. The same formula described an ideal gas in equilibrium, which he had checked out in Planck's textbook on thermodynamics and which Jakob confirmed. If humanity's economic behavior was the same as the behavior of atoms of a gas, an exact national economy and correct social reform seemed within reach. The two professors relighted their cigars while they took up the philosophical question of whether humanity was material or gas mental.

As long as one could still have conversations like these . . . Looking into the smoky flame he forgot where he was for a moment. He was alone in the town square. His cigar finished, he started home but then turned in the direction of the physical institute. Before he had gone far, he turned again and found himself back in the center of the square. There, more or less facing the statue of Bismarck, he began to speak. His voice sounded clear and confident.

If it is science that lifts a corner of the veil of ignorance, it is love and fidelity that bind soul to soul. The truest words I know are inscribed on Kant's tomb: The starry heavens above me and the moral law within me. In that order, I intended to talk to you about the laws that unite nature and about the moral laws that unite us. I intended to ask you first to join with me in a walk in the luminous heights of pure physical research, which is distant for many of you . . . But I realize now that, with all true Germans, your thoughts are occupied with the war and the future of our country.

What can I tell you then? I am a physicist.

This year or next I will retire after fifty years, since a younger man is needed for the good of the physical institute. It will be hard for me to leave the institute, which I've loved and made sacrifices for. The new man, I know, will spend his days calculating spectra from the Bohr atom and not worry about the old physics of the world-ether, and that is how it must be. My life in physics began with Maxwell's equations as the exact equations of the ether, and if certain hypotheses prove right it may end with them. And it began with the German Reich and it may end with that, too. My first meeting of the German Association of Natural Scientists and Physicians was a celebration of the new Reich after the defeat of France. People there said that German science had aided Germany's rise to power, that scientists had kept alive the beautiful ideal of German unity, that Germany's victory in the field was a victory of the German mind. Before and after 1870, in our nation's painful rise to the place that its industrious people deserved and its neighbors denied, physicists like other good Germans willingly sacrificed their personal interest to the larger interest.

Do I believe that the international spirit of science is evidence that in science love is stronger than hate? Yes. Even in the midst of this endlessly destructive war I believe that science is a bearer of understanding between peoples, a solvent of national hatreds, a reagent of love. In our innermost beings, we science professors are researchers (even though I mainly teach these days), and as researchers we are eminently peaceful, unworldly, international. Who among us hasn't passed back and forth across national frontiers in the name of his science? I think of our colleagues and friends, of Marie Curie in Paris, of Rutherford in Manchester, of Joffé in St. Petersburg, and I think of the tragedy of this ruinous war that scientists didn't want.

Don't misunderstand me. We support Germany in the war, even though we know it may bring a temporary end to relations with our colleagues abroad and will certainly damage our science. It is no betrayal of our scientific creed for us to yield to the dominant pull of patriotism, despite what scientists abroad may say about us. To work for the fatherland, we temporarily sacrifice what is essential to our work as scientists, but we do not for a moment renounce our desire for a peaceful world that knows no national scientific barriers. By its nature science differs from other work. In technology and business, people work against each other; in art, they work side by side; in science and only in science, they work seriously together, in harmony instead of in conflict, in selfless coordination instead of in egotistic isolation.

At this moment the political world does not follow our example. Abroad they abuse Planck, Wien, and other colleagues for declaring their solidarity with the German army, as if the army hadn't prevented the destruction of Germany and its culture in the past, as if the army weren't part of the German people. In our hearts we feel as patriotic Germans, which doesn't prevent us from honoring an intellectual and moral world beyond the struggle between nations.

From the Netherlands, a neutral country, Lorentz has written to us about common concerns, about the growing hatred between peoples and the interruption of scientific exchanges—and about the mistreatment of the Belgians by the German army. If Lorentz, the clearest head in all of theoretical physics, can believe the terrible things that they say about us, there is little reason to expect an understanding of our position anywhere. (Lorentz understands us better now.) When the Prussian Academy moved to expel foreign members—Ramsay and scientists of the French Academy—Planck urged successfully against it because it would have disastrous consequences for academies everywhere and so for international science. He knew that to move against the eventual return of international relations was to sin against science.

The truth is that inner contradictions of national cultures have been hidden by a common technical and scientific culture and that these contradictions now militate against the best traditions of science. If Einstein understands that, does he also understand that our hearts are torn? (Einstein really isn't so far from us now.)

But I am not a political man. I have signed none of the endless manifestos my colleagues have circulated. If our universities had armies, our manifestos might mean something. Cannon alone speak meaningfully for the German cause, and it doesn't become scientists and scholars to pretend to participate in the war by lending their names to high-sounding pronouncements. They write manifestos, while at the front teachers from my seminar hurl themselves on the barbed wire.

I dread opening each new issue of the *Physikalische Zeitschrift* or of the *Berichte* of the German Physical Society. I dread the black-bordered notices of fallen German physicists. My colleagues at the front behave heroically, if it makes sense to talk of heroes who are gassed, cannoned, mortared, mined, bombed, machine-gunned, and grenaded. The theoretical physics professors Hasenöhrl and Cantor volunteered and were killed in the beautiful Tirol when Italy fell on the back of Austria. In the fullness of his scientific powers, Hasenöhrl was killed by a grenade while leading his company. Cantor was killed by a mine in the mountains, where he was driven by the thought that his highest duty was to fight for the fatherland. Hasenöhrl was immersed in all the modern developments and had a brilliant future in the strange paths that theoretical physics was certain to follow. Cantor couldn't follow the paths leading beyond classical physics into relativity and quantum theories. Wien put it well when he said that Cantor

Sign. 2452, Manuscript Collection, Deutsches Museum, Munich.

p. 50 *The job had nothing to do with that:* A friend believed that Drude's fall in the mountains resulted in a mental disturbance and that this contributed to his death. Hans Falkenhagen, "Zum 100. Geburtstag von Paul Karl Ludwig Drude (1863–1906)," *Forschungen und Fortschritte*, 37 (1963), 220–221.

p. 50 At the end of a drinking party given for him in Marburg, Althoff gave a speech in which he likened professors to whores. Kayser, "Erinnerungen," p. 172.

The Institute

p. 52 In the event that the Emperor toured the physical institute and noticed the bad conditions there, Althoff could then assure him that plans for a new institute were ready. Kayser, "Erinnerungen," p. 253.

p. 53 The description of Jakob's new institute largely follows that of Bonn's new physical institute in 1913. H. Kayser and P. Eversheim, "Das physikalische Institut der Universität Bonn," *Physikalische Zeitschrift*, 14 (1913), 1001–08.

p. 53 Hertz compared confidential enquiries with official calls. Letter to A. Schulte, 20 April 1889, Sign. 2761, Manuscript Department, Bonn University Library.

p. 54 In his disappointment, Jakob acted like Auerbach, who quieted his nerves in Munich after the Jena faculty opposed his promotion from extraordinary to ordinary professor for theoretical physics. Letters from Elisabeth Foerster-Nietzsche to Anna Auerbach, 28 December 1911, and from Harry Federley to Felix

Auerbach, 26 January 1912, Auerbach Papers, Prussian State Library.

p. 54 In proposing the title and rank of honorary ordinary professor for Jakob, his faculty gave reasons similar to the Tübingen faculty's for proposing the same for their extraordinary professor for theoretical physics Karl Waitz: his long, diligent, and notable services as representative of theoretical physics and his temporary services as representative of experimental physics. Minister of Church and Education to the King of Württemberg, 28 June 1907, Sign. E 14, Bü 1608, Main State Archive, Stuttgart.

p. 57 *Most of them came from abroad:* Voigt reported that many foreign students worked in his institute for theoretical physics in Göttingen and that for the most part they came highly recommended and were good. He recalled the time when there were no German institutes like his in which students could do research, and so German students had to go to Paris and London, where they were liberally admitted to research institutes. Germany now had an honorable duty to receive foreign students in the same spirit. Letter to Curator Osterrath, 21 March 1912, Sign. XVI. IV. C. v, Curatorial File, Göttingen University Archive.

p. 57 James Franck recalled how he and fellow students all but lived in the Berlin physical institute, drawn together as a family by Warburg's demand for total commitment. "Emil Warburg zum Gedächtnis," *Die Naturwissenschaften*, 19 (1931), 993–997.

p. 57 Kayser described how the Bonn physical institute emptied out at the start of the war ("Erinnerungen," pp. 295–296), a story that was repeated in other physical institutes throughout Germany. Although enrollments fell off sharply, they remained large enough to strain the reduced teaching resources. At Leipzig, for example, Wiener was surprised to find the enrollment in his lecture course still a third of its normal size and in his practical course still a quarter. Letter to Wien, 24 November 1914, Manuscript Collection, Deutsches Museum, Munich.

p. 58 In *Grosse Männer* (Leipzig: Akademische Verlagsgesellschaft, 1909) Wilhelm Ostwald characterized Helmholtz as a classical type, on p. 260, and on pp. 371–388 he developed his general distinction between this and the romantic type.

p. 59 Des Coudres was identified as one of Ostwald's romantic scientists by Wiener, "Nachruf auf Theodor Des Coudres," *Berichte über die Verhandlungen der sächsischen Akademie der Wissenschaften zu Leipzig. Mathematisch-physikalische Klasse,* 78 (1926), 358–370.

p. 59 *On colorful stationery:* Des Coudres wrote such letters and cards to Wiener from exotic places before World War I. Wiener Papers, Manuscript Department, Karl Marx University Library, Leipzig.

p. 59 In the Wiener Papers and in Des Coudres' personnel file in the Archive of Karl Marx University, Leipzig, there are many letters from September and October 1914 dealing with Des Coudres' volunteer service. The ministry recognized Des Coudres' willingness to sacrifice, but it ordered him back from the field to teach thermodynamics and to make up for lost time by lecturing every day.

p. 60 *At the time, he was furious with protesters:* A draft, dated 21 March 1916, of a protest by Wien against the protesters and correspondence pertaining to it are in the Wien Papers, Prussian State Library, Berlin.

p. 61 Critical of the lack of organization of scientific talent, Wien regretted that German physicists had no greater significance in the war. Discussion on pp. 36–37 and letter to Gustav Mie, 21 August 1916, on p. 63 in Wien, *Aus dem Leben und Wirken eines Physikers.*

p. 61 Wien explained his plans for a postwar liaison between the military and university scientific institutes in one of his lectures to the army in the Baltic. *Vorträge über die neue Entwicklung der Physik und ihrer Anwendungen. Gehalten im Baltenland im Frühjahr 1918 auf Veranlassung des Oberkommandos der achten Armee* (Leipzig: J. A. Barth, 1919), p. 110.

p. 62 *Jakob missed the Russian assistant:* Voigt's reasons for asking his foreign assistants to resign. Letter to the Curator, 10 November 1914, Sign. 4/V h/35, Göttingen University Archive.

p. 63 Kohlrausch cautioned Wien about an assistant who had joined a Catholic society. People with strong Catholic faith, Kohlrausch explained, lost their independence and ability to reason after a certain point. Letter of 13 March 1899, Sign. 2436, Manuscript Collection, Deutsches Museum, Munich.

p. 66 Bonn replaced its extraordinary professor for theoretical physics Hermann Lorberg, who was committed temporarily to a mental institution. Insisting that he was able to resume his lectures when he was released, Lorberg did not recognize the need for his replacement. Allowed to go on lecturing, Lorberg continued his practice of filling the enormous blackboard with formulas, which he would refer to on succeeding days. His replacement, however, lectured in the same hall, only at a different hour, and he needed the blackboard as well. Lorberg was enraged to find his formulas erased one day, and he promptly wrote them all back again. Kayser, "Erinnerungen," pp. 232–233; letter by Lorberg to Curator von Rottenburg, 7 January 1903, Sign. IV E II b, Lorberg, Bonn University Archive.

p. 67 Jakob's colleagues who sang or played instruments included Helmholtz, who brought to his acoustics lecture the harmonium he designed for playing either the pure or the well-tempered scale. In front of the class one day, he improvised skillfully on both scales, as a quarter of an hour passed, then a half hour. Still he played on, and only when restless students got up noisily did he stop to look around, surprised. By contrast, Helmholtz' student Hertz was tone-deaf, more so than Kayser, his neighbor in the Berlin physical institute, believed possible. Lost in thought, Hertz paced his room late at night singing impure tones, which lacked all melody and caused Kayser much misery. From time to time, Kayser asked Hertz to stop. Kayser, "Erinnerungen," pp. 106, 139.

p. 71 Des Coudres sent many postcards from the field to Wiener, who wrote back that his Leipzig circle was interested in hearing of his direct experience of the war. Letters from Wiener to Des Coudres, 23 October 1914 and 16 August 1915, Wiener Papers, Manuscript Department, Karl Marx University Library, Leipzig.

p. 72 Between 1915 and 1917, the Jena publisher Gustav Fischer brought out four editions of *Die Physik im Kriege.* Auerbach wrote the book on the assumption that people who could not fight in the war nevertheless wanted to experience it richly.

p. 73 Wien explained to his publisher B. G. Teubner that his time had been taken up with U-boat work. Letter of 22 July 1918, Sign. P-12, reel 3, Archive for History of Quantum Physics, American Philosophical Society Library, Philadelphia.

p. 73 *He found it hard to put himself into a physicist's frame of mind:* Auerbach supposed that people who were not at the front were startled by an inner voice, which reminded them that whatever they were doing was senseless. *Die Physik im Kriege*, 3rd ed., p. 1

p. 73 The Geheimrath's difficulty with Bohr's atomic model was widely shared at this time. Now convinced that it gave the correct representation of spectroscopic facts, Carl Runge was still troubled. The explanation that radiation occurred when an electron passed from one path to another seemed to him all too fanciful, and he asked Sommerfeld what, after all, was real understanding in physics. Letter during the war, quoted in Iris Runge, *Carl Runge und sein wissenschaftliches Werk*, p. 167.

p. 74 Jakob's views on hypotheses largely agree with the Königsberg theoretical physics professor Paul Volkmann's in his *Erkenntnistheoretische Grundzüge der Naturwissenschaften und ihre Beziehungen zum Geistesleben der Gegenwart* (Leipzig: B. G. Teubner, 1896), pp. 61–62.

p. 74 *At the end of Jakob's passionate speech, the Geheimrath looked unhappy:* The Geheimrath's petulant thoughts about theoretical physicists were shared by Johannes Stark, *Die gegenwärtige Krisis in der deutschen Physik* (Leipzig: J. A. Barth, 1922), pp. 1–2.

p. 76 A custodian in the Leipzig physical institute tormented Wiener until, in desperation, Wiener put the case before the ministry. He said that not only had the custodian lied, betrayed, and resisted him, but he had slandered a young woman in his household and tried to force his way into his maid's bedroom. Wiener now spoke to the custodian only with a witness present, and in general he was so shaken that he had fallen behind in his research and in his administration of the institute. He couldn't begin the semester while the custodian was still there, and when the custodian was finally removed Wiener requested a week's leave with his doctor's support. After the strain of dealing with the custodian, Wiener explained, he needed a rest. Letters by Wiener to the Saxon Ministry of Culture and Public Education, 30 March 1903, and to Dean K. Bücher, 17 April 1903, Wiener's personnel file, Karl Marx University Archive, Leipzig.

p. 77 At Bonn the theoretical physics lecturer Bucherer claimed that valuable apparatus of his had been stolen from a storage room in the institute. There was good reason to suspect the former custodian's son, who had stolen from the institute before and who had been forbidden to enter the institute, even when his parents still lived there. Bucherer regarded the director of the institute Kayser as responsible for his losses and demanded compensation. The University Curator and Judge decided against Bucherer's claim. Letters to Curator Ebbinghaus by A. H. Bucherer, 8, 19, and 29 November 1916; by H. Kayser, 10 and 22 November 1916; by P. Eversheim, 10 November 1916; and by the University Judge, 13 November 1916, Bonn University Archive.

p. 77 The second time Lorberg found his formulas erased, he reserved a small square for his replacement's formulas and filled up the rest of the board again with his own. Kayser, "Erinnerungen," p. 233.

p. 77 When Lorberg failed to get Kayser to grant him control of the blackboard, he threatened to get his revolver and lie in wait ("Erinnerungen," p. 234). Lorberg regarded Kayser as an enemy who wanted him put away and who told students he was crazy. Kayser took Lorberg's threat on his life seriously, but Lorberg's physicians believed him harmless and only senile. Although Lorberg had recently bought cartridges, he had owned the revolver for years, as the Bonn police reported to the mayor; they were not much worried that he might use it. Lorberg's health and competence as a lecturer in theoretical physics and Kayser's concern to keep Lorberg out of his institute are the subject of letters and reports from 1901 to 1906 by Lorberg, Kayser, the Bonn Curator, the Mayor of Bonn, Althoff in the Prussian Ministry of Culture, and the physicians Wirsch and Pelmann, Sign. IV E II b, Bonn University Archive.

p. 77 Early in the world war, Kayser became interested in the problem of locating enemy guns by acoustical means. Finding a solution, he built an apparatus and conducted experiments with pistol shots in the garden of the Bonn physical institute. He reported his results to the war ministry and got permission to take his apparatus to the western front to try it out in major battle. It proved useless. Kayser, "Erinnerungen," p. 301.

p. 78 *Indeed, modern Germany was inconceivable without it:* From Wiener's argument that the natural sciences and technology were part of culture, in fact its foundation. *Physik und Kulturentwicklung,* p. 100.

p. 80 *They took him up high over the town:* After giving a course for flyers at Leipzig University, Wiener went up in a plane, an experience he described to Des Coudres. Letter of 16 August 1915, Wiener Papers, Manuscript Department, Karl Marx University Library, Leipzig.

p. 80 *The Geheimrath recalled that before the war:* From Wiener's reflections on the theme of war and peace in his *Vogelflug, Luftfahrt und Zukunft* (Leipzig: J. A. Barth, 1911), pp. 21–24, 36–46.

p. 81 *Without warning:* On the road to the airfield, the car Wiener was riding in turned over, an accident he related to Des Coudres. Letter of 27 July 1915, Wiener Papers, Manuscript Department, Karl Marx University Library, Leipzig.

p. 82 Volkmann's views on the deterioration of research and teaching owing to pressures on today's physicists. *Franz Neumann. 11. September 1798, 23. Mai 1895. Ein Beitrag zur Geschichte deutscher Wissenschaft* (Leipzig: B. G. Teubner, 1896), pp. 27–28.

p. 83 Jakob shares Voigt's guarded acceptance of the technical disciplines in universities, which Voigt spoke about in his welcoming address to the Göttingen Union for the Promotion of Applied Physics and Mathematics ("Ansprache Seiner Magnificenz des Herrn Prorectors," 1911/12, Voigt Papers, Manuscript Department, Göttingen University Library). To Voigt it was an article of faith that the physicist's first responsibility is to strive for knowledge, which demands selflessness and devotion. He assumed that the useful application of knowledge has so many enticements that research in it will arise by itself every time. *Physikalische Forschung und Lehre in Deutschland während der letzten hundert Jahre. Festrede im Namen der Georg-August-Universität zur Jahresfeier der Universität am 5. Juni 1912* (Göttingen, 1912), p. 18.

p. 83 Jakob speaks out against the denigration of science in the name of art as Volkmann did in *Erkenntnistheoretische Grundzüge der Naturwissenschaften*, 2nd ed. (Leipzig: B. G. Teubner, 1910), pp. 290–293.

p. 86 Jakob's typical letter was one Hertz received from a physician, who traced erroneous views of phenomena to the division of the natural sciences, which were no science at all because they were only parts of science. Letter from H. Struve, 13 August 1890, Sign. 3057, Manuscript Collection, Deutsches Museum, Munich.

p. 87 Within an institute building, the experimental and theoretical professors usually had harmonious enough relations, but there are a good many instances of difficult relations like Jakob and the Geheimrath's here. At Halle, Schmidt and Dorn had a falling out, at Heidelberg Pockels and Lenard, at Freiburg Königsberger and Himstedt, and elsewhere others. Directors were constantly on their guard against anyone's assertion of rights in teaching or research in their institutes. In turn, those who lived under a director's rule felt oppressed at times. Bucherer, for example, believed that Kayser was the destroyer of his career and the cause of inadequate teaching of theoretical physics at Bonn. Bucherer's response was to box up his things and quit his room in Kayser's institute.

It was the middle of World War I, and the familiar tensions within the physical institute were charged with patriotic feeling. Letters in 1916 and 1917 by Bucherer, Kayser, the Bonn Curator, and an official of the Prussian Ministry of Culture, Curatorial and Philosophical Faculty Files, Bonn University Archive.

The conflict between Röntgen and Leo Graetz at Munich was almost inevitable. As extraordinary professor for theoretical physics, Graetz had acquired an unintended independence in the physical institute during the illness and following the death of Röntgen's predecessor. With Röntgen's arrival as director, misunderstandings arose, and the faculty, senate, and ministry all were drawn in. Eventually Graetz was removed as director of the student laboratory. In recognition of his long service to the university—and of the hopelessness of his position there, for after Sommerfeld's arrival his teaching in theoretical physics also became superfluous—he was elevated to honorary ordinary professor in 1908. Graetz formally applied at this time for a place for his teaching and research in Röntgen's institute, which Röntgen replied to with an eleven-page letter. The rules it contained for the use of the instruments came down to the need for one man, Röntgen, to make all decisions. Röntgen needed all work room in the institute for students and for instruments that were now spilling into the corridor, so there was no space for Graetz's research. There could be no question of a student laboratory under Graetz's direction since that would only lead to collisions. Graetz could occasionally call on the porter, and occasionally a student of Graetz's could work in the institute. That was all. His thirty years' experience had taught him how to run an institute, Röntgen said. Graetz complained that he couldn't be expected to ask Röntgen's permission every time he needed a piece of equipment, which would put him on the level with a beginning student. But Röntgen's mind was made up. Letters by Graetz of 6 and 24 May and by Röntgen of 20 May and 5 June 1908, Sign. OC-N 14, Munich University Archive.

p. 88 The Geheimrath's two schemes for winning the war were not beyond the imagination of the time. Rutherford received a scheme for attacking submarines with vortex rings (letter from H. T. Barnes, 22 September 1915, Rutherford Papers, Sign. Add. 7653/B32, Cambridge University Library), and J. J. Thomson received one concerning the existence of matter repelled by gravity, which the Board of Inventions and Research wanted back (letter from G. H. Brady, 17 April 1918, J. J. Thomson Papers, Sign. Add. 7654(i)/B69, ibid.).

p. 89 To Runge, who had pointed out an error in a paper of his, the Zeiss physicist S. Czapski wrote in detail about his illness and the pressure of work. He believed that for these reasons he was free from reproach for error. Letter of 21 January 1893, Sign. 1948/55, Manuscript Collection, Deutsches Museum, Munich.

p. 90 Evoking the blood of youth, Dionysian mystery, beauty, joy, and ecstasy in the face of the dead facts of science, Gottfried Benn's 1914 Expressionist play *Ithaka* portrayed an old German professor of pathology who deserved to be murdered by his assistant and students for valuing science above everything else. Benn, *Gesammelte Werke*, ed. D. Wellershoff, vol. 2 (Wiesbaden: Limes Verlag, 1958), pp. 293–303.

p. 91 Lorberg told Kayser that if he weren't so old, he would gladly fight a duel with him. Letter by Kayser to Curator von Rottenburg, 27 June 1903, Sign. IV E II b, Lorberg, Bonn University Archive.

The World-City

p. 93 Drude began a correspondence with Hertz by asking him about the electrical prism he used to confirm Maxwell's theory. Drude described his own work on the metallic reflection of light, which Hertz encouraged. Hertz expected further understanding of the new electromagnetic theory from the side of optics, in which he did not regard himself an expert as he did Drude. Letters from Hertz to Drude, 30 April, 15 May, 4 August 1892, and 6 February 1893, Sign. 3208, 3209, 3210, and 3212, Manuscript Collection, Deutsches Museum, Munich.

p. 94 In a letter to Wiener, Drude spoke of his generally poor luck in the matter of calls. He brought up the call to Hannover that hadn't materialized five years before, and he referred to Wiener's confidential remarks on the ministerial official involved, who must have been Althoff. Letter of 12 March 1899, Wiener Papers, Manuscript Department, Karl Marx University Library, Leipzig.

p. 94 *To top this, Althoff spread rumors:* Drude described his negotiations with Althoff over Hannover and Kiel in a series of letters to Kayser, who was keeping him informed on the preferences of the Hannover faculty (letters of 1, 4, 7 June, 7 July, and 1 August 1894, Darmstädter Collection, Sign. F 1 c 1897(5), Prussian State Library, Berlin). Drude had learned that Althoff was talking about him; Drude gave the impression, according to Althoff, that he felt superior to Hannover. He was innocent of that, he told

Kayser. He would gladly have gone to Hannover (letter of 10 August 1894, ibid.).

p. 94 To introduce Maxwell's theory of electricity and magnetism in an easily understood way, Drude wrote his first treatise, *Physik des Aethers auf elektromagnetischer Grundlage* (Stuttgart: F. Enke, 1894). His inaugural lecture at Leipzig in 1894 appeared as *Die Theorie in der Physik* (Leipzig: S. Hirzel, 1895).

p. 95 In the event of a Leipzig call, Drude said, he would be glad that he would not have to negotiate with Althoff but only with Saxon officials whom he knew well (letter to Wiener, 8 October 1902, Wiener Papers, Manuscript Department, Karl Marx University Library, Leipzig). When the call came, he did not accept it, but he went out of his way to praise the Saxon officials for their openness throughout the negotiations (letter to Wiener, 4 November 1902, ibid.).

p. 96 Notices in daily papers gave details of the commemoration; for example, "Trauerfeier für Prof. Drude," *Tägliche Rundschau,* 7 July 1906, and "Professor Dr. Drude," *Berliner Tageblatt,* 8 July 1906.

p. 97 Personal details from Drude's letters to Wiener of 30 December 1899, 8 April 1900, 22 June 1900, 8 October 1902, 15 March 1903, 12 June 1903, 2 August 1903, and from Drude's wife Emilie's letter to Wiener, 30 May 1901 (Wiener Papers, Manuscript Department, Karl Marx University Library, Leipzig), and from Drude's letter to Sommerfeld on 13 January 1901 (Sommerfeld Papers, Manuscript Collection, Deutsches Museum, Munich).

p. 97 Drude told of physicists falling in the mountains in a letter to Wiener, 8 June 1900, Wiener Papers, ibid.

p. 98 *Jakob now held a letter from Switzerland:* Drude described this unhappy time to his close friend Richard Lorenz. Letter of 6 November 1905, Darmstädter Collection, Sign. F.1 c 1897(5), Prussian State Library, Berlin.

p. 100 When Drude died, the revision of his optical textbook was halfway through publication, and his assistants saw the rest of it through. Drude discussed Planck's quantum theory of black-body radiation on pp. vii–viii, 512–519, and Einstein's theory of relativity on pp. 467–468 *Lehrbuch der Optik*, 2nd ed. (Leipzig: S. Hirzel, 1906).

p. 102 Planck had advised Drude on the worthlessness of Bucherer's paper, as he told Wien. Letters of 29 November and 7 December 1906, Wien Papers, Sign. Akz. Nr. 1973.110, Prussian State Library, Berlin.

p. 103 Upset by a paper published in the *Annalen der Physik* without his knowledge, Planck talked to Drude about it. Drude agreed that the paper was bad, but the author had appealed to him personally and he hadn't the heart to refuse. Drude had to deal with the resulting polemics. Letter from Planck to Wien, 12 October 1906, ibid.

p. 103 *Although the physical world picture had been shaken:* Jakob's use of an architectural image for theoretical physics was familiar. For example, Max von Laue, *Das physikalische Weltbild. Vortrag gehalten auf der Kieler Herbstwoche 1921* (Karlsruhe: C. F. Müllersche Hofbuchhandlung, 1921), p. 6.

p. 104 *It wouldn't surprise him if the relativity and quantum theories:* This was Planck's judgment on what was most significant in physics over the previous decade. Conceding that physics used to be simpler, more harmonious, more trustworthy, and generally more satisfying, Planck said that physics could not overlook the new developments if it were not to come to a standstill. Letter to Wien, 13 June 1922, Wien Papers, Sign. Akz. Nr. 1973.110, Prussian State Library, Berlin.

p. 104 Sommerfeld recognized the boldness of the philosophical mind of the nation of poets and thinkers in the relativity and quantum theories of Einstein and Planck. With this claim no nation could quarrel, Sommerfeld told the German Women's Union of the Red Cross for the Colonies in April 1918. "Die Entwicklung der Physik in Deutschland seit Heinrich Hertz," *Deutsche Revue*, 43 (1918), no. 3, 122–132.

p. 105 *Die Naturwissenschaften* devoted an entire issue in 1918 to the papers given in honor of Planck's sixtieth birthday.

p. 108 *Everyone who served the demanding goddess of science:* Ostwald's thoughts about the inevitable sadness in the life of a devoted scientist, in "Gustav Wiedemann," *Berichte über die Verhandlungen der königlich sächsischen Gesellschaft der Wissenschaften zu Leipzig. Mathematisch-physikalische Classe*, 51 (1899), lxxvii–lxxxiii. Elsewhere he applied these thoughts to Boltzmann and Drude (*Grosse Männer*, pp. 401–402).

p. 108 *Jakob remembered that in his forties he first became aware:* As Richard Willstätter recalled the loss of freshness and happiness in his scientific work in *Aus meinem Leben*, p. 282.

p. 109 Jakob's diary resembled Hertz's, which showed an emergent interest in physics embedded in other interests (*Erinnerungen, Briefe, Tagebücher*, pp. 24–34).

Jakob's diary resembled Franz Neumann's, too, in its concern with controlling the self through scientific study (Luise Neumann, *Franz Neumann*, pp. 76–78). Otherwise, Neumann's diary belonged to a much earlier time than Jakob's, and its religious preoccupation had a different intensity from his. Jakob's youthful religious questioning had a rational origin in a contradiction he saw between scriptures and science, and in this respect his diary resembled Des Coudres'. Troubled by the contradiction, Des Coudres began keeping a diary at the time of his confirmation, as Wiener reported in "Nachruf auf Theodor Des Coudres," p. 365.

p. 110 *Instead of settling for either pure or applied physics:* Shortly before Drude's first appointment, at Leipzig, the Göttingen physicists and mathematicians wanted Drude made extraordinary professor. They pointed to his success in teaching electrotechnology, and they wanted him to continue in this field he had won for himself and the university. Joint letter by E. Schering et al. to Minister Bosse, 7 November 1893, Drude's personnel file, Sign. 4/V c/205, Göttingen University Archive.

p. 110 An example of Drude's tendency to undervalue his work was an entry in a questionnaire for an album that his university was assembling. Drude wrote that he was editor of the *Annalen der Physik* and then he crossed it out, as if on second thought deciding it was unworthy of mention. Questionnaire from the Rector to Drude, 7 May 1900, Drude's personnel file, Giessen University Archive.

p. 110 Only a few days after announcing Drude's election to the Prussian Academy of Sciences, the *Vossische Zei-*

tung on 6 July 1906 announced his death the day before. The notice reported that Drude's closest friends had long been concerned that he might break down under the weight of work and the mental strain. Similar but shorter notices were published that day in other newspapers such as the *Berliner Tageblatt* and *Tägliche Rundschau.* Of the many full obituaries of Drude, Planck's memorial address to the German Physical Society is the most revealing; published in the society's proceedings, the address "Paul Drude" is reprinted in Planck's *Physikalische Abhandlungen und Vorträge* (Braunschweig: F. Vieweg, 1958), III, 289–320.

p. 111 *How could this happen:* The Heidelberg physics professor Georg Quincke's reaction was widespread—with Drude's death, he said, one stood before a riddle. Letter to J. W. Brühl, 10 July 1906, Sign. Heidelberg Hs. 3632,8, Manuscript Department, Heidelberg University Library.

p. 111 *You didn't call out:* Drude never asked for help, and so Planck and others concluded he didn't need or wish it. Drude could have unburdened himself of certain duties, which made it all the more puzzling to Planck. Letter to Wien, 30 July 1906, Wien Papers, Sign. Akz. Nr. 1973.110, Prussian State Library, Berlin.

p. 112 Drude's close friend Wien expressed his sorrow in a poem in which he said that Drude's friends stayed quiet, didn't hear him, and now were stunned and could only ask why he did it. *Aus dem Leben und Wirken eines Physikers,* pp. 24–25.

p. 113 In 1905 and early 1906, during Drude's year in Berlin, doubts were raised about certain calculations of his on the role of the earth in wireless telegraphy. If the doubts were sound, one of Drude's assistants had just published a totally false result. To lay the doubts to rest, Drude wanted to do the necessary experiments himself. Other claims on his time so frustrated his project that it wasn't until the middle of June 1906 that he was able to set up his experiments in the institute garden. His death three weeks later prevented him from finishing them. From Drude's diary and journals and with his widow's permission, his assistant published the results of the experiments. "Beeinflussung einer Gegenkapazität durch Annäherung an Erde oder andere Leiter," *Annalen der Physik*, 21 (1906), 123–130.

p. 113 *Of course, overwork alone:* Kohlrausch's opinion on the catastrophe in a letter to Wien, 15 September 1906, Sign. 2452, Manuscript Collection, Deutsches Museum, Munich.

p. 114 To Jakob as to Kohlrausch, a university the size of Berlin's was a monster, and he didn't envy an institute director in such a place. Letter to Wien, 15 September 1906, Sign. 2452, Manuscript Collection, Deutsches Museum, Munich.

p. 116 *On the other hand, the old saying that the work of an institute:* Jakob is thinking in particular of Freiburg's tiny institute for theoretical physics, where Königsberger directed experimental research as best he could. Lacking funds for assistants, tables and chairs, and even such an elementary tool as a lathe, he paid out of his own pocket to keep it going. Letters in 1904 and after between Königsberger, the Philosophical Faculty, and the Baden Ministry of Justice, Culture, and Education, Sign. G.L.A. 235, No. 7769, General State Archive, Karlsruhe. The poor conditions in Freiburg had come to Jakob's attention just before the war, since Königsberger had sent out a questionnaire to theoretical physicists asking about the arrangements in their institutes. Jakob realized that his own arrangements were good, relatively speaking; for the two rooms Königsberger had been using for his doctoral students' research had been taken away from him, leaving his institute with a single room to serve at once as work space, director's office, and a place for apparatus. Letter from Königsberger to Auerbach, 8 January 1913, Auerbach Papers, Prussian State Library.

p. 116 *Jakob easily remembered back:* Regrets at the passing of the old ways of doing physical research from Jakob's senior colleague A. Paalzow, "Stiftungsfeier am 4. Januar 1896," *Verhandlungen der deutschen physikalischen Gesellschaft*, 15 (1896), 36–37.

p. 117 *What was given to the big was taken from the small:* Voigt believed that German physics would lose its standing if smaller institutes continued to be neglected and bigger ones favored. He had in mind in particular the giant Imperial Institute of Physics and Technology in Berlin. *Physikalische Forschung*, pp. 16, 20–21.

p. 118 Kayser, too, learned about the writer Gottfried Keller from his Strassburg physics professor August Kundt. "Erinnerungen," pp. 73–74.

p. 120 The researching physicist was often likened to a mountain climber, an image Wien carried through in detail. *Aus dem Leben und Wirken eines Physikers*, pp. 49–50.

p. 120 Elisabeth Abbe described her husband Ernst's restless activity to Anna Auerbach, 3 October 1897, Auerbach Papers, Prussian State Library, Berlin.

The World

p. 123 Jakob visited Italy and shared Kayser's original delight in the beauty of country and city and his later despair over the encroachments of modern life. On revisiting San Remo, Kayser found the wild flowers gone and a modern spa with a gas factory. On revisiting—in his imagination—a Greek temple near Salerno, he found the holy solitude and poetry destroyed and the smell of gas in the air. "Erinnerungen," pp. 12–16, 20–22, 47–53.

p. 124 Jakob's trip to Greece repeated Wien's trip, as Wien described it in letters to his wife in 1912. Wien admired the view of the Acropolis and Athens from Colonus, Sophocles' birthplace (*Aus dem Leben und Wirken eines Physikers*, pp. 54–60). For the classically educated man, according to Kayser, the Acropolis brought together the landscape, the architecture, and the idea of the glory of Greece. On his first visit, his father took him to Colonus, where he told him the story of Sophocles' Oedipus. That became a favorite of Kayser's, and he read it often, in Greek. He later had a chance to see *Antigone* performed in Greek in the Stadium, with the Acropolis in view. Greece, Kayser said, was like walking in a dream ("Erinnerungen," pp. 122–127 and passim).

p. 125 *He had taken an examination:* Long after graduating from secondary school, Kayser often dreamed of the final examination. For his university examination he had been better prepared, but he was faulted for his ignorance of Latin syntax. Later he took part in an examination of a physics student whom the examining philologist asked to discuss the laws of Athens in a given year, believing that everyone including physicists ought to know all about that. "Erinnerungen," pp. 98–101.

p. 131 They heard Walter Hasenclever's five-act tragedy *Antigone* (Berlin: P. Cassirer, 1917). It was first performed at the end of 1917, despite tight censorship. It won the Kleist Prize that year and was widely reviewed.

p. 133 *The Classical Physicist knew it might be fanciful, but:* Carl Neumann's comparison of physics to classics in "Worte zum Gedächtniss an Wilhelm Hankel," *Berichte über die Verhandlungen der königlich sächsischen Gesellschaft der Wissenschaften zu Leipzig. Mathematisch-physikalische Classe,* 51 (1899), lvii–lxiii.

p. 134 Jakob's views on the closely related themes of abstraction and the world-ether in physical theory were prompted by a long debate that continued even after 1918. The tenor of his thinking on these themes can be gathered from a sample of the popular or semitechnical writings he would have read with interest, if not always with agreement.

The physicist's "cold gray cave of abstraction" was Mie's expression. From that cave came the abstract, imperceptible world-ether, which conferred unity on the physical world picture (*Die Materie. Vortrag gehalten am 27. Januar 1912* [*Kaisers Geburtstag*] *in der Aula der Universität Greifswald* [Stuttgart: F. Enke, 1912], pp. 6–7, 20, 32). Relativity theory had replaced the theory of the world-ether for some physicists but not for Emil Wiechert, who believed that relativity theory in fact provided new indications of the existence of the world-ether ("Relativitätsprinzip und Äther," *Physikalische Zeitschrift,* 12 [1911], 690).

Voigt discussed the mathematical approach of relativity theory and the light-quantum hypothesis in connection with the "modern movement to dispense with the ether hypothesis" and predicted that for now the gain in visualizability would continue to support the ether hypothesis ("Strahlende Aetherenergie," *Handwörterbuch der Naturwissenschaften*, ed. E. Teichmann et al., vol. 9 [Jena: G. Fischer, 1913], 769). For Auerbach, the alternative to the visualizable world picture based on the world-ether was one in which energy alone existed; because this world picture lost everything in abstraction, he thought that most physicists felt dissatisfied with it (*Das Wesen der Materie, nach dem neuesten Stande unserer Kenntnisse und Auffassungen dargestellt* [Leipzig, Dürr'sche Buchhandlung, 1918], pp. 142–144). The inevitability of abstraction in physics was insisted on by Emil Cohn; in general relativity theory, he pointed out, the evaporation of space and time into mathematical concepts wholly removed from their origin in sense perception was only another instance of the abstraction that every new comprehensive law imposed on physics (*Physikalisches über Raum und Zeit*, 3rd ed. [Leipzig and Berlin: B. G. Teubner, 1918], pp. 26–28).

Jakob had been provoked by Einstein's observation a few years back that the ether belonged to a point of view that physics had overcome. So had Lenard been provoked, but Jakob saw no value in Lenard's way of answering Einstein, which was to resurrect a truly superseded standpoint. With a mechanically conceived ether, Lenard argued, physicists had a good physical picture to work with. They didn't have one if they rested content with mathematical formulas (*Über Relativitätsprinzip*, *Äther*, *Gravitation* [Leipzig: S. Hirzel, 1918], pp. 10–12). The "sick man of theoretical physics," the ether, wasn't dead, Einstein replied to Lenard. General relativity theory endowed empty space with physical properties, and so it was permissible to speak of an ether as long as it didn't have material properties ("Dialog über Einwände gegen die Relativitätstheorie," *Die Naturwissenschaften*, 6 [1918], 701–702).

Since the eclipse expedition in 1919, Einstein's general relativity theory had been the empirically correct physical world picture, according to Richard Von Mises (*Naturwissenschaft und Technik der Gegenwart* [Leipzig and Berlin: B. G. Teubner, 1922], p. 15). According to Johannes Stark, however, general relativity had less to do with physics than with mathematics and philosophy. Physics without the ether was no physics, in his view; the ether was a fact, and the visual thinking it promoted had proved fruitful in physics (*Die gegenwärtige Krisis in der deutschen Physik*, p. 12).

In holding to the world-ether hypothesis, Jakob belonged to the company of "many excellent physicists," as Nernst put it. As one himself, Nernst attributed an energy to the ether at absolute zero temperature, and from it arose all the matter of the universe (*Das Weltgebäude im Lichte der neueren Forschung* [Berlin: J. Springer, 1921], p. 32). Other of Jakob's colleagues also insisted on the importance of the ether. Wiechert, for one, argued that the laws of relativity theory were a consequence of the properties of the ether and that all forces were borne by the ether and that all matter arose from it ("Der Äther im Weltbild der Physik," *Nachrichten von der königlichen Gesellschaft der Wissenschaften zu Göttingen. Mathematisch-physikalische Klasse*

[1921], pp. 30–36). For another, Mie argued that Einstein's gravitational theory belonged to a physical world picture based on a unified world substance, the ether (*Die Einsteinsche Gravitationstheorie. Versuch einer allgemein verständlichen Darstellung der Theorie* [Leipzig: S. Hirzel, 1921], pp. 27–33). Jakob noted that Sommerfeld, the supporter and developer of the new atomic theory, had no quarrel with the word *ether.* It was sufficient for his purposes to regard the ether as an infinite system of immaterial oscillators (*Atombau und Spektrallinien* [Braunschweig: F. Vieweg, 1919], pp. 247–248).

If Jakob had lived on after the war, he might have written an essay on the world-ether in light of recent developments in physics. More likely, he would have quietly waited for experimental facts or astronomical observations to restore confidence in the physics of the ether, as his colleague Max Abraham did (as Born and von Laue report in "Max Abraham," *Physikalische Zeitschrift,* 24 [1923], 52). In any case, Jakob would have recognized his own point of view in Graetz's lectures. The question of the ether was the leading question of physics, Graetz wrote. He foresaw the continued use of visualizable pictures in physics in addition to the abstract, purely conceptual approach of relativity theory, and he invited physicists to bring relativity and quantum theories and their views of the atomic constitution of matter to the question of the ether and to undertake new researches to fathom its nature (*Der Äther und die Relativitätstheorie. Sechs Vorträge* [Stuttgart: J. Engelhorns Nachf., 1923], pp. 1, 54, 65, 80).

p. 134 *He had recorded the gathering political clouds:* From Voigt's recollections of his conversion from physics student to soldier. *Erinnerungsblätter aus dem deutsch-französischen Kriege 1870/71*, pp. 9, 11–15, 44, 96.

p. 136 *How did it happen, Helmholtz demanded:* To be overlooked in the assignment of windows was not a trifling matter for the new director. Letter to Du Bois-Reymond, 17 March 1871, Darmstädter Collection, Sign. F 1 a 1847, Prussian State Library, Berlin.

p. 137 Voigt recalled a sad soldiers' song and the gala celebration at the Brandenburg Gate. *Erinnerungsblätter aus dem deutsch-französischen Kriege 1870/71*, pp. 16, 210.

p. 137 *What would become of that unhappy nation France:* Helmholtz had served as a physician in the war and even visited a battlefield right after the battle. His son was in the war and had been hurt. Letters to Du Bois-Reymond, 17 October 1870 and 14 February 1871, Darmstädter Collection, Sign. F 1 a 1847, Prussian State Library, Berlin.

p. 137 Helmholtz' arrangements for Kayser to do research in the old Berlin institute and Kundt's for Kayser to do research in the new Strassburg institute. "Erinnerungen," pp. 71–74, 93–94.

p. 138 *Before he left, he saw colossal steam engines ram great posts:* From Anna von Helmholtz' description of the new Berlin physical institute under construction. The new institute that Helmholtz was promised when he came to Berlin was delayed for years. The responsibility for that, his wife said, lay with the many important men in the ministries. Letter to Leo Königsberger, n.d., Darmstädter Collection, Sign. F 1 a 1847, Prussian State Library, Berlin.

p. 139 *Planck had returned a thoughtful criticism:* Hermann Weyl's new gravitational theory was an example of researches that sought to subsume all phenomena under a world formula, here Hamilton's principle. Planck thought that such researches were bound to increase and that they were potentially valuable even if they could never fully realize their goal. Letter of 16 September 1915, Wien Papers, Sign. Akz. Nr. 1973.110, Prussian State Library, Berlin.

p. 141 *Soon after the war began, he received a letter:* From a letter filled with hatred toward the English, written early in the war by a well-known young German theoretical physicist. Klein, *Paul Ehrenfest*, p. 301.

p. 142 In an address at Cambridge in 1912, Kayser said that he didn't feel like a stranger, that Germans and Englishmen were kin. "Erinnerungen," pp. 285–289.

p. 142 The offending English physicist was Oliver Lodge, as Cohn reported to Hertz. Letters of 8 June and 22 August 1889, Sign. 2881 and 2882, Manuscript Collection, Deutsches Museum, Munich.

p. 143 The printed manifesto signed by sixteen German physicists mentioned a statement hostile to Germans signed by English scholars and scientists, among them Lodge, Lord Rayleigh, J. J. Thomson, and other physicists. A copy of the German manifesto is in the Felix Klein Papers, Sign. III, A, Manuscript Department, Göttingen University Library.

p. 143 *On reflection, Jakob disapproved of the manifesto:* Jakob's reasons for disapproving were the same as Planck's. Letter to Wien, 1 January 1915, Wien Papers, Sign. Akz. Nr. 1973.110, Prussian State Library, Berlin.

p. 143 After the war started, German and British colleagues continued to correspond for a while. Hans Geiger wrote to Rutherford as late as spring 1915 to report on German physicists like himself who were serving at the front. Letter of 26 March 1915, Rutherford Papers, Sign. Add. 7653/G52, Cambridge University Library.

p. 143 The interned physicist was Peter Pringsheim, who asked his teacher Rutherford to send him papers so that he could keep in touch with science. Letter of 15 November 1915, P86, ibid.

p. 144 While interned, Chadwick wrote to Rutherford several times and mentioned Geiger, Warburg, and other physicists. Rutherford was early informed about Chadwick's teaching in the camp by H. Robinson, 17 July 1915, R44, ibid.

p. 144 Geiger's understanding of Chadwick's internment as atonement followed from his belief that England bore all guilt for the war. On this point, Geiger was quoted by A. F. Kovarik. Letter to Rutherford in early 1915, K55, ibid.

p. 144 After seeing Planck and Warburg in Berlin, the Dutch theoretical physicist H. A. Lorentz reported to Rutherford that they did not subscribe to the idea of Germany's superiority over other countries. Letter of 18 August 1915, L134, ibid.

p. 145 In his diary the young Göttingen mathematician Richard Courant wrote at the outbreak of war that it was a wonderful dream. Reid, *Courant*, p. 49.

p. 145 Planck valued the political unity the war had brought to Germany and believed it would last, but he was sorry that it had to come about this way. Letter to Wien, 8 November 1914, Wien Papers, Sign. Akz. Nr. 1973.110, Prussian State Library, Berlin.

p. 145 *Just the other day he heard from a young mathematical physicist:* Courant wrote to Hilbert about his desire to be out of uniform and back in science and about the stupidity of the intellectuals during the war. Letter quoted in Reid, *Courant*, p. 71.

p. 146 Wien explained why relativity theory in physics did not imply a general relativity of judgments. *Die Relativitätstheorie vom Standpunkte der Physik und Erkenntnislehre* (Leipzig: J. A. Barth, 1921), pp. 25–26.

p. 146 The economic behavior of humanity was compared to the physical behavior of an ideal gas by Auerbach's friend Walter Harburger. Letter, n.d., Auerbach Papers, Prussian State Library, Berlin.

p. 147 *But I realize now that, with all true Germans:* Röntgen's observation on the hold of the war over the thoughts and feelings of Germans. Letter to M. O. Boveri, 27 April 1918, quoted in Nitske, *The Life of Wilhelm Conrad Röntgen*, p. 250.

p. 147 To the Curator at Göttingen, Voigt explained his readiness to step down to make way for a younger man. But it would be hard for him. Letter of 29 March 1915, Voigt personnel file, Sign. 4/V b/203, Göttingen University Archive.

p. 148 Like Voigt, Jakob recognizes the opposing pulls of national patriotism and scientific internationalism. But Jakob stops short of Voigt's identification of German national interests with mankind's international ones. "Ansprache gelegentlich der Zusammenkunft der Lehrer der Georgia-Augusta am 31. Oktober 1914."

p. 149 *Abroad they abuse Planck:* Jakob here recalls criticism from abroad of the physicists who were among the ninety-three German scholars, scientists, and artists who signed the August 1914 manifesto "To the Cultural World."

p. 149 *In our hearts we feel as patriotic Germans:* Planck regretted the wrong impression that the manifesto of the ninety-three had given rise to, and in an open letter to Lorentz he affirmed his continuing belief in international cultural goods. Lorentz published Planck's letter in the Netherlands and circulated copies to colleagues. Letter of March 1916, Felix Klein Papers, Sign. III, A, Manuscript Department, Göttingen University Library.

p. 149 *From the Netherlands, a neutral country, Lorentz has written:* Lorentz wrote to Wien, hardly knowing why except that he had the feeling Wien should know. Letter of 22 March 1915, Sign. 2481, Manuscript Department, Deutsches Museum, Munich.

p. 149 Wien had taken pains to answer Lorentz' charges about the conduct of the German army, which he believed were based entirely on Lorentz' feelings instead of on cases. Letter to Planck, 1 May 1915, Wien Papers, Sign. Akz. Nr. 1973.110, Prussian State Library, Berlin.

p. 149 *Lorentz understands us better now:* Although his sympathies lay with the other side, Lorentz conceded to Planck that the German position could be represented. Letter from Planck to Wien, 4 May 1915, ibid.

p. 149 *He knew that to move against the eventual return:* After the war Planck wrote, poignantly, of the continuing breakdown of international relations in science: "Today this time of a trusting international communal life lies behind us like a beautiful, long vanished dream." "Gedächtnisrede des Hrn. Planck auf Heinrich Rubens," in *Sitzungsberichte der preussischen Akademie der Wissenschaften zu Berlin. Physikalisch-mathematische Klasse, 1923*, p. cxii.

p. 150 *If Einstein understands that:* Einstein and three others at the University of Berlin answered the manifesto of the ninety-three with a "Manifesto to Europeans," which among other things criticized scientists and artists for yielding to nationalist passions instead of rallying to preserve a common world culture. Later Einstein joined Planck in signing a proposal for early compromise with England. He had growing sympathy for the suffering Germans.

p. 150 Not manifestos but cannon alone speak for Germany, Planck said, and the scientists and scholars should withdraw. Letter to Wien, 1 January 1915, Wien Papers, Sign. Akz. Nr. 1973.110, Prussian State Library, Berlin.

p. 150 At the beginning of the war, the editors of the *Physikalische Zeitschrift* requested directors of physical institutes to keep them informed about the war activities of their men. This policy not only expressed gratitude to the defenders of Germany, they said, but informed foreigners that German physics supported the fatherland in its need and danger. Born to Wien, 23 November 1914, Manuscript Collection, Deutsches Museum, Munich.

p. 150 Cantor and Hasenöhrl were both Austrian, though Cantor taught theoretical physics in Germany and Hasenöhrl in Austria. In Jakob's theoretical physics, there was remarkably little change in teaching personnel in the German universities throughout the war. What change there was came about in the usual ways, mainly through natural death. The main exception was Cantor.

p. 151 *Even physics cannot always help:* Einstein wrote to Lorentz of his constant depression, from which he could no longer escape by turning to physics. Letter of 18 December 1917, published in *Einstein on Peace*, p. 21).

The Mountains

p. 154 *He began climbing again:* During vacations, Lorberg liked to make difficult first climbs in the Dolomites. He also liked to climb in the more inaccessible parts of the local mountains, though this was child's play by comparison. In his last weeks he took to climbing in the local mountains at night. One morning he was found dead after having fallen into a ditch half full of water at the foot of one of the mountains. His feet were stuck fast in the clay bottom, while his head and shoulders were above water. Kayser, "Erinnerungen," p. 234; letter by the Bonn Curator to the Prussian Minister of Culture, 6 March 1906, Sign. IV E II b, Bonn University Archive.

p. 155 *He saw crowds moving through the streets:* From Werner Heisenberg's dream of the streets around Munich University in 1919. *Physics and Beyond: Encounters and Conversations*, trans. A. J. Pomerans (New York: Harper and Row, 1972), p. 125.

ACKNOWLEDGMENTS

For their critical reading of this book, I wish to thank Martin J. Klein and Thomas S. Kuhn. For their encouragement in developing the approach of this book, I wish to thank Robert H. Kargon and Michael A. Aronson.

My research was done with the help of a grant from the National Science Foundation.

I am still resisting with all my might the "meaninglessness" of the events of the world and trying to replace it by "incomprehensibility." But how difficult it is to carry through this point of view.

—Max Planck to Arnold Sommerfeld, 15 December 1919